Build Your Own 3-Axis CNC Router Machine

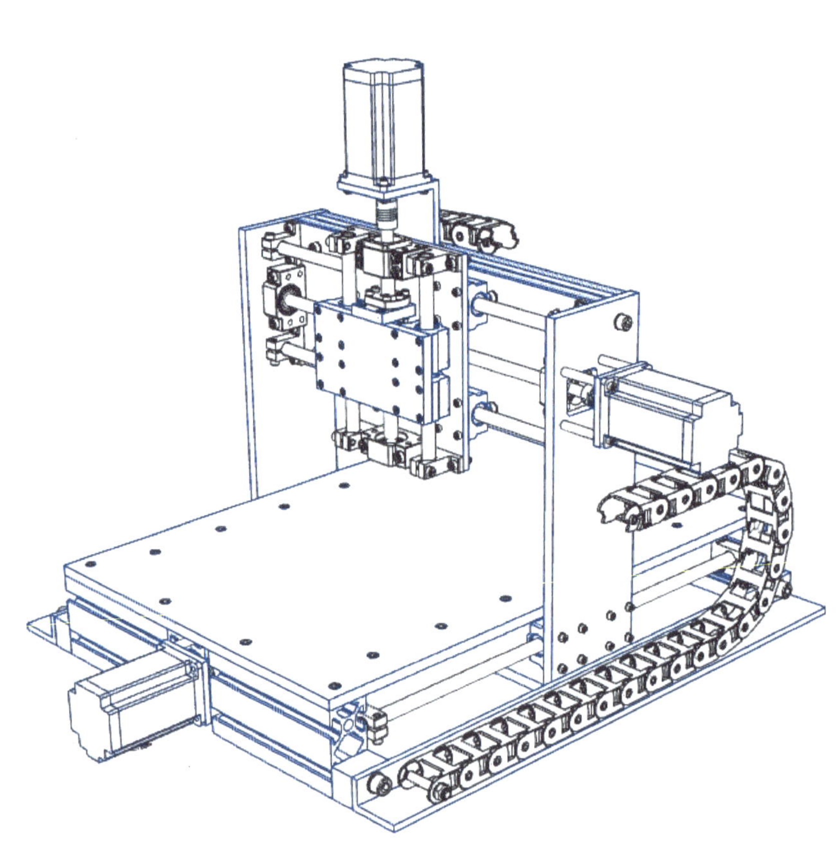

Build Your Own 3-Axis CNC Router Machine

Tsipas Ilias

2015

First Printing: 2015

ISBN 978-960-93-7188-9

Lemnos, Greece

Contents

Preface

The latest years CNC (Computer Numerical Control) machines are considered as the most innovative projects in the field of technology. Nowadays it is easier for each home to have a CNC Router Machine in order to design and build prototypes parts.

I hope that you will find this document useful and at the same time entertaining as you build your CNC Router Machine.

The machine has been designed in a way so that the average builder can construct the machine with common electrical tools, low budget and minimum working hours.

If you need advice with the construction, don't hesitate to contact me on my email: ilias86gr@gmail.com

Machine Specifications

Machine Size:

Overall Machine size: 690 x 570 x 550mm

Exact usable Z axis travel will depend on exact configuration of spindle mount, length of bits and thickness of the spoiler board if one is used.

Dimensioning:

Plans are dimensioned in Metric measure. They are not available in Imperial but a conversion on the dimensions can be done from builder.

Drive Method:

X Axis: Ball-screw with Anti-Backlash nut
Y Axis: Ball-screw with Anti-Backlash nut
Z Axis: Ball-screw with Anti-Backlash nut

Spindle:

The machine can use a simple trim router or a 2.2 – 3.0KW spindle (ER11 to ER20) with VFD.

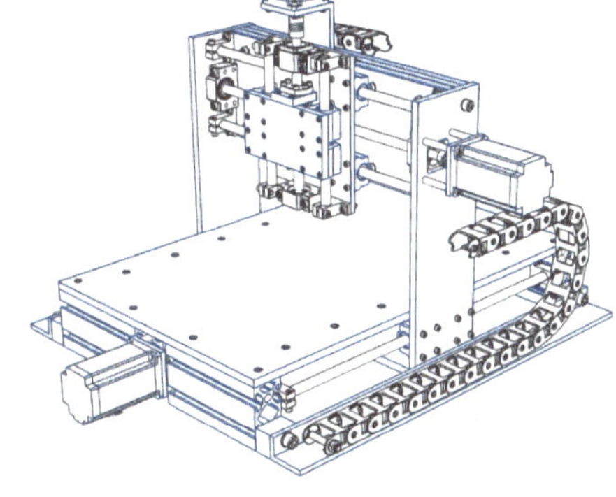

Stepper Motors:

NEMA 23 frame size.
Minimum recommended size: **275oz/in**

General Notes

- ✓ Do not measure off of printed drawings. Use dimensions as indicated on drawings. Scale is not indicated on drawings due to variation in printer accuracy. Printing at 100% factor may not guarantee exactly 100% on paper.
- ✓ Do not make changes in the design without fully understanding of their implications.
- ✓ Dimensions on mechanical parts are given in metric decimal units. Dimensions are given to either 2 decimal place accuracy or may be given as full decimal equivalents to fractions. This does not indicate degree of tolerance required.
- ✓ Tolerances. An accuracy of +/- 0.03mm is generally sufficient on metal parts. Any exceptions to this will be noted.
- ✓ **WORK SAFE.** Use good judgment while working and do not attempt anything that is beyond your ability or that may jeopardise your personal safety.

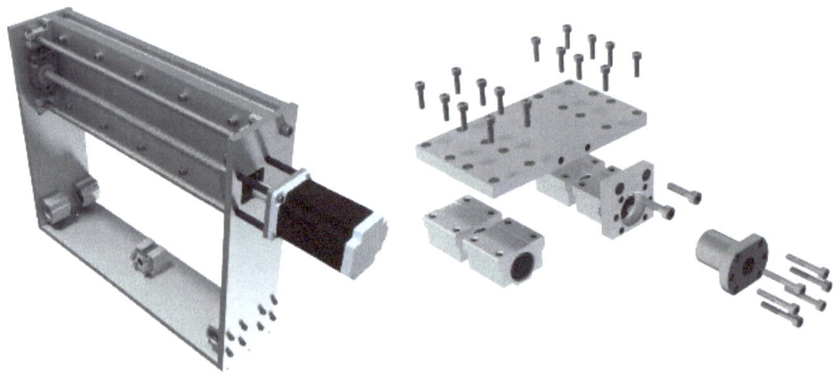

Desktop Manufacturing

What was once a technology that existed strictly within industry, CNC equipment has increasingly found widespread use in the home workshop. At its most basic, CNC is a method of using a numerical code to control a machine. Nearly any type of machine or configuration can be controlled this way. If it has a range of movement, whether linear or rotary, it can be controlled by a numerical code that instructs those movements. Therefore, CNC can be used on a wide spectrum of equipment, such as milling machines, lathes, plasma cutters, water jets, hot wire foam cutters, wire EDM or, as is the case with these plans, a *3-Axis Router Machine*.

In the early development of CNC, the number that were used to control a machine were hand-coded and punched into a paper roll that was fed through a mechanical reader. The punched holes equated to discrete movement steps. While programming simple movements, such as straight lines, was easily accomplished, curved or free-from geometry was much more difficult to achieve. With these complex shapes, the smaller the distance between the motion steps, the smoother the results, necessitating the calculation of thousands of movement points. With the advent of the computer came the ability to generate much more complex numerical code, resulting in very smooth machine motions.

More recently there has been a tremendous growth in DIY home-built CNC equipment. It is now a relative straightforward process to generate the *G-Code* to control a CNC machine tool on a home PC and output the signal through a parallel port or USB port to motion control motors. Depending on how complex and sophisticated the geometry of the parts being manufactured, this can be accomplished with free software.

The potential implications for this revolution in "desktop manufacturing" are huge. Transferring the manufacture of extremely complex parts from costly industrial settings, which was the only option in the recent past, to a low-cost home shop opens up a new world of possibilities.

However, having the ability to easily have these manufacturing capacities available raises the questions of appropriate use of technology. If a part can adequately be made by more traditional hand crafted methods, it may be an inefficient use of time and resources to use a computer controlled machine. In addition to the initial time invested in machine construction, the fabrication of a part can require significant time spent at the computer. Even a simple part requires drawing or 3D modelling it, deciding on a machine strategy, generating toolpaths and G-Code from the drawing and setting up the stock to be cut in the machine.

The advantages comes in using the equipment for purposes that cannot be achieved easily by other means. CNC lends itself to jobs requiring high levels of accuracy, consistency between repetitive parts and cutting complex geometries. These advantages are significant and potentially transformative, for both hobby and business use.

Bill of Materials 1

	Assembly	Material	Dimensions (mm)
Aluminum	Carriage Table	6061 - T6 Flat	144x80x10
			40x40x10
	Z-Axis	6061 - T6 Flat	200x144x7
		6061-T6 Angle	82x66x6.35 @60
		6061-T6 C-Shape	90x40x6.35 @60
	Y-Axis	6061 - T6 Flat	2x 340x108x10
		6061 - T6 Flat	420x108x10
		6061 - T6 Flat	400x144x5
		6061 - T6 Flat	40x30x5
		6061-T6 Angle	50x30x6.35 @400
		Profile 40x40	2x @400
	X-Axis	6061 - T6 Flat	580x380x10
		6061 - T6 Flat	2x 500x50x15
		6061 - T6 Flat	40x25x30
		6061 - T6 Flat	60x60x35
		6061-T6 Angle	2x 32x60x6.35 @100
		6061-T6 Angle	2x 50x30x6.35 @500
		Profile 80x40	2x @160
		Profile 80x40	@380
Steel	Y-Axis	4140	4x ø8x2.5 @42
	X-Axis	4140	4x ø8x2.5 @42

Table showing the rough dimensions for the parts need to be fabricated. More trimming required for the parts to be ready for the sub-assemblies placement.

Bill of Materials 2

Assembly	Quantity and Size	Hardware
Carriage Table	(16) M4x14	Socket Head Cap Screws
	(6) M4x25	
	(2) M5x20	
	(6) M4	Prevailing Torque Nuts
	(22) M4	Plain Washers
	(2) M5	
Z-Axis	(4) M3X16	Socket Head Cap Screws
	(26) M4X16	
	(12)M5X20	
	(6)M6X50	
	(4) M4	Prevailing Torque Nuts
	(8) M5	
	(6) M6	
	(4) M3	Plain Washers
	(26) M4	
	(12)M5	
	(6)M6	
Y-Axis	(4) M3x10	Socket Head Cap Screws
	(22) M4x20	
	(12) M5x20	
	(6) M5x25	
	(6) M5x50	
	(10) M6x16	
	(2) M6x20	
	(4) M8x45	
	(2) M8x60	
	(6) M6	Prevailing Torque Nuts
	(4) M3	Plain Washers
	(22) M4	
	(24) M5	
	(12) M6	
	(6) M8	
X-Axis	(12) M5X20	Socket Head Cap Screws
	(8) M6x10	
	(8) M6x16	
	(12) M6x20	
	(6) M6x40	
	(2) M8x70	
	(4) M10x50	

X-Axis	(8) M5	Prevailing Torque Nuts
	(12) M5	Plain Washers
	(34) M6	
	(2) M8	
	(4) M10	

Table showing the hardware needed for the sub-assemblies.

List of Fabricated Parts

	Part	Part Name	Qty	Material
Carriage Table	1.1	Carriage Table	1	Aluminum
	1.2	Ball-Nut Mount	1	Aluminum
Z-Axis	2.1	Z-Axis Back Plate	1	Aluminum
	2.2	Stepper NEMA23 Mount	1	Aluminum
Y-Axis	3.1	Z-Axis Back Plate	1	Aluminum
	3.2	Profile 40x40	2	Aluminum
	3.3	Right Side Plate	1	Aluminum
	3.4	Left Side Plate	1	Aluminum
	3.5	Lower Plate	1	Aluminum
	3.8	Cable Bracket	1	Aluminum
	3.11	NEMA 23 Mount Spacer	4	Steel
X-Axis	4.1	Base Plate	1	Aluminum
	4.2	Table Side Plate	2	Aluminum
	4.3	Profile 80x40	2	Aluminum
	4.4	Profile 80x40	1	Aluminum
	4.5	BK Mount	1	Aluminum
	4.6	BF Mount	1	Aluminum
	4.7	Cable Bracket	2	Aluminum
	4.8	Cable Bracket Mount	4	Aluminum
	4.11	NEMA 23 Mount Spacer	4	Steel

Table showing the parts need to be fabricated.

List of Bought Parts

	Part	Part Name	Qty
Carriage	1.3	SC 12	4
Table	1.3	SFU 12	1
Z-Axis	2.3	Rail ø12mm and L:300mm	2
	2.3	Rail Mount SK12	4
	2.3	Flexible Coupling Shaft	1
	2.3	Leadscrew ø12mm L:224mm	1
	2.3	BF 10 Bearing Mount	1
	2.3	BK 10 Bearing Mount	1
Y-Axis	3.6	Rail ø12mm and L:400mm	1
	3.6	Rail Mount SK12	4
	3.6	Flexible Coupling Shaft	1
	3.6	Leadscrew ø12mm L:420mm	1
	3.6	BF 10 Bearing Mount	1
	3.6	BK 10 Bearing Mount	1
	3.7	SC 12	4
	3.7	SFU 12	1
	3.7	Ball-Nut Mount for 1204 BS	1
X-Axis	4.9	Rail ø12mm and L:500mm	2
	4.9	Rail Mount SK12	4
	4.9	Flexible Coupling Shaft	1
	4.9	Leadscrew ø12mm L:507mm	1
	4.9	BF 10 Bearing Mount	1
	4.9	BK 10 Bearing Mount	1

Table showing the parts need to be bought.

Assemblies - Exploded Views

Carriage Table Assembly – Exploded View

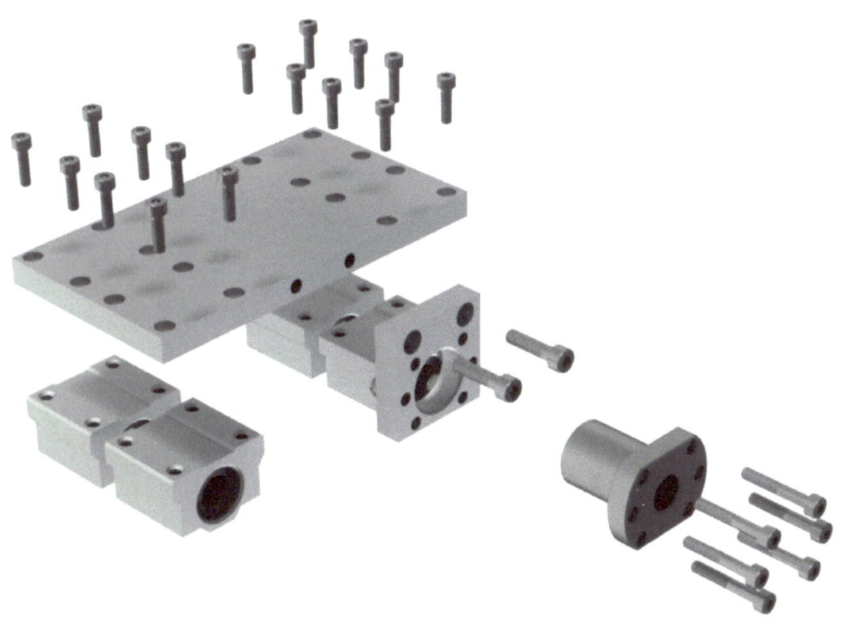

Z-Axis Assembly – Exploded View

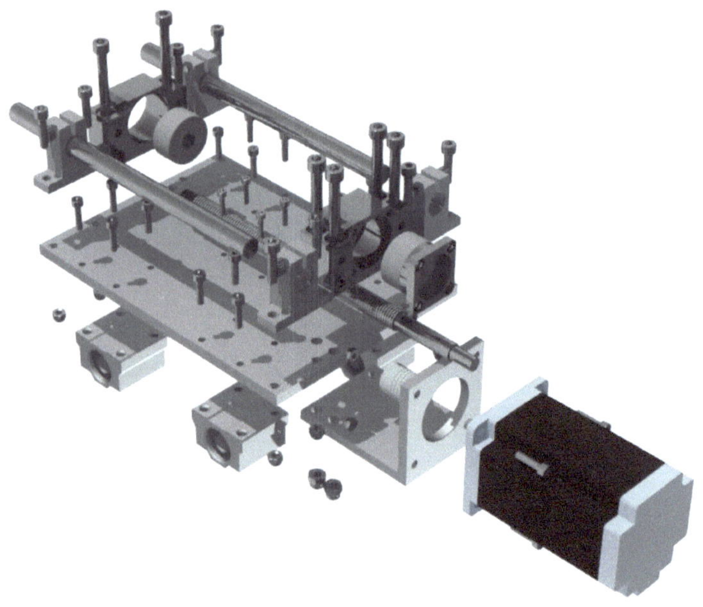

Y-Axis Assembly – Exploded View

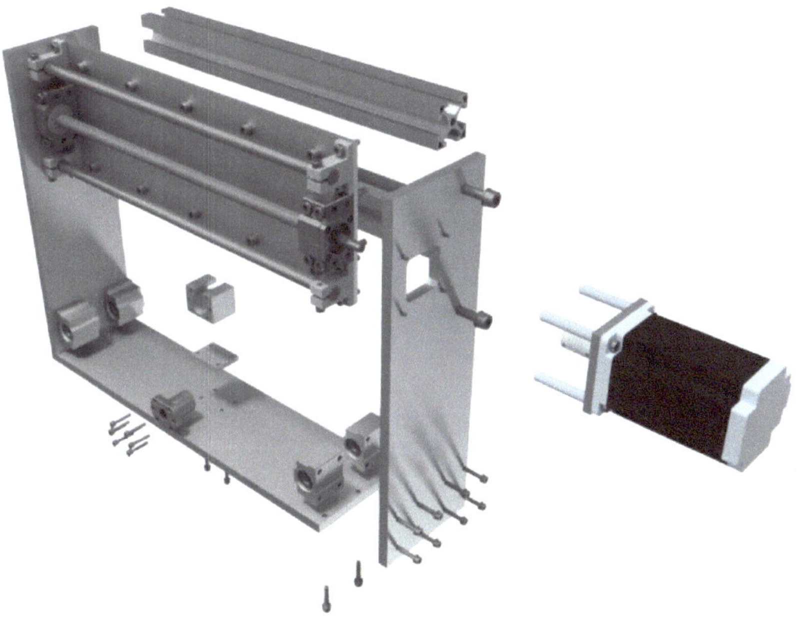

X-Axis Assembly – Exploded View

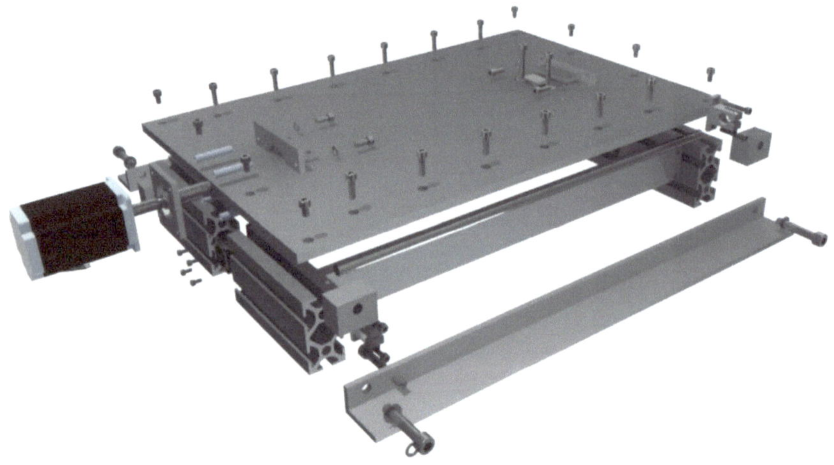

Drawings

In the following pages, the drawings for the parts need to be manufactured are listed. **All parts are made of aluminum.**

While making those parts, don't print the drawings in order to stick them on the raw material stocks but just use the dimensions to lay out the drawings on the raw material.

If you will use an angle grinder to cut the parts, always cut "outside of the line" and then file the parts to the desired dimension. If you can't be 100% accurate, it is better to have a part slightly bigger than smaller.

Most of the parts have chamfers on the edges but that is something that does not have to be done if you don't have the right tools to make it. It will not affect the working condition of the machine at all.

In some drawings like the aluminum profiles that need only to be cut in length there is no need for a full dimension drawing. It is provided only the dimension needed to cut (length dimension).

Enjoy the build and always work safe!

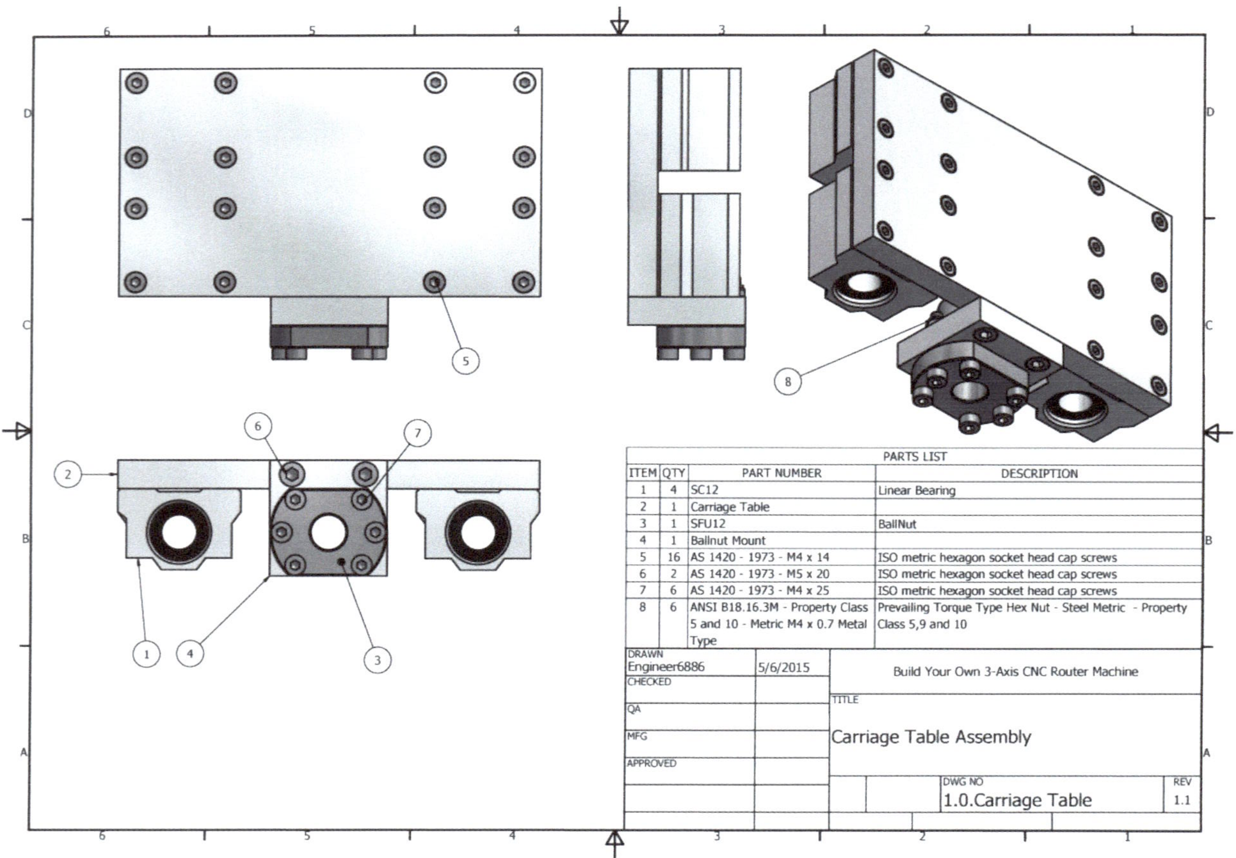
PARTS LIST

ITEM	QTY	PART NUMBER	DESCRIPTION
1	4	SC12	Linear Bearing
2	1	Carriage Table	
3	1	SFU12	BallNut
4	1	Ballnut Mount	
5	16	AS 1420 - 1973 - M4 x 14	ISO metric hexagon socket head cap screws
6	2	AS 1420 - 1973 - M5 x 20	ISO metric hexagon socket head cap screws
7	6	AS 1420 - 1973 - M4 x 25	ISO metric hexagon socket head cap screws
8	6	ANSI B18.16.3M - Property Class 5 and 10 - Metric M4 x 0.7 Metal Type	Prevailing Torque Type Hex Nut - Steel Metric - Property Class 5,9 and 10

DRAWN
Engineer6886
5/6/2015
CHECKED
QA
MFG
APPROVED
Build Your Own 3-Axis CNC Router Machine
TITLE
Carriage Table Assembly
DWG NO
1.0.Carriage Table
REV
1.1

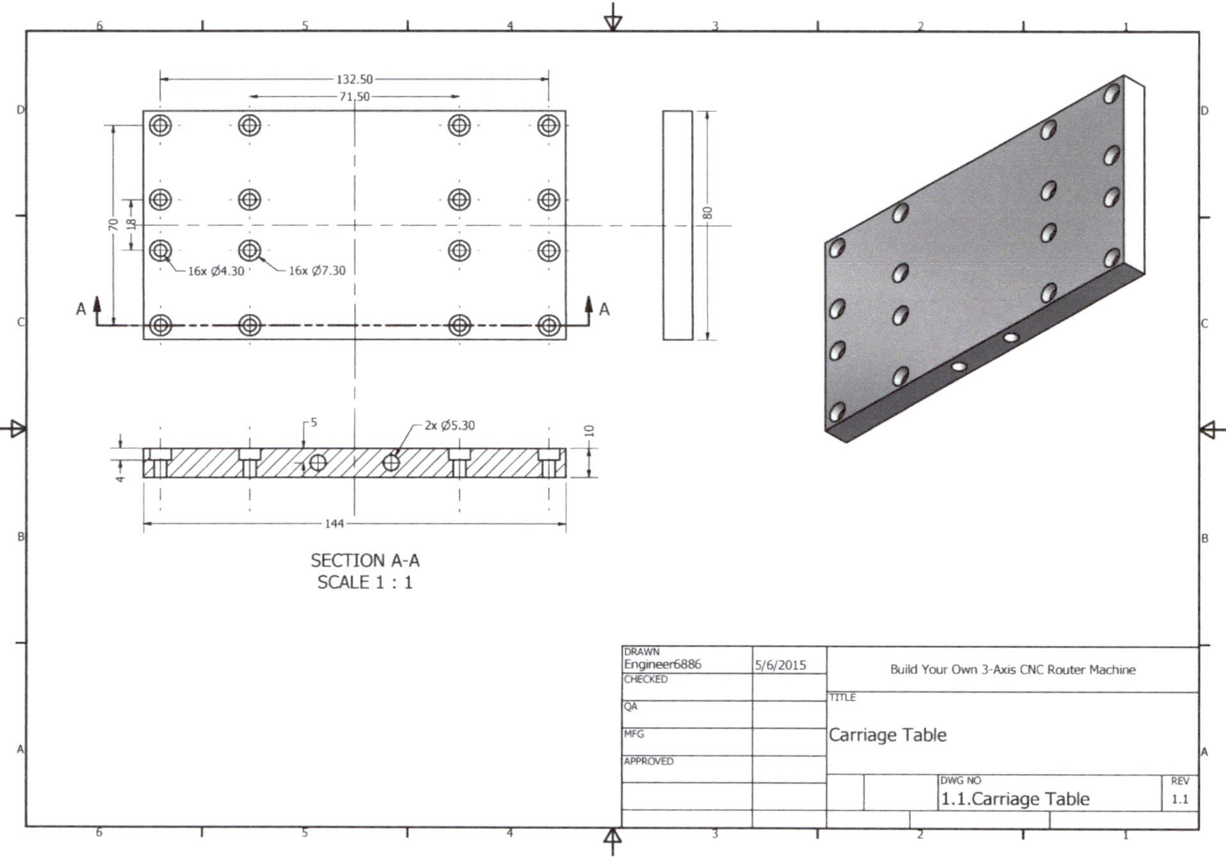

132.50
71.50
70
18
16x Ø4.30
16x Ø7.30
A
A
80
5
2x Ø5.30
10
4
144
SECTION A-A
SCALE 1 : 1
DRAWN
Engineer6886
5/6/2015
CHECKED
QA
MFG
APPROVED
Build Your Own 3-Axis CNC Router Machine
TITLE
Carriage Table
DWG NO
1.1.Carriage Table
REV
1.1

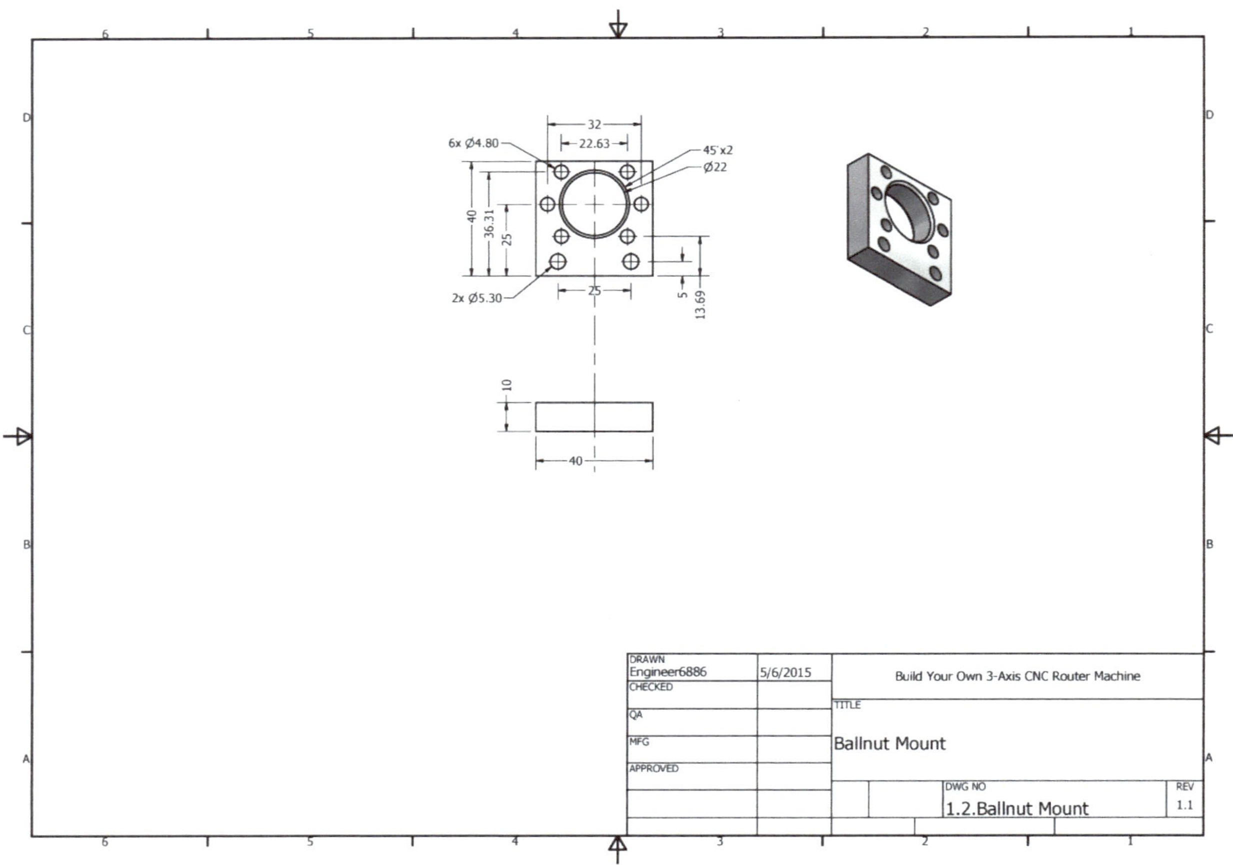
6x Ø4.80
32
22.63
45°x2
Ø22
40
36.31
25
25
5
13.69
2x Ø5.30
10
40
DRAWN
Engineer6886
5/6/2015
CHECKED
QA
MFG
APPROVED
Build Your Own 3-Axis CNC Router Machine
TITLE
Ballnut Mount
DWG NO
1.2.Ballnut Mount
REV
1.1

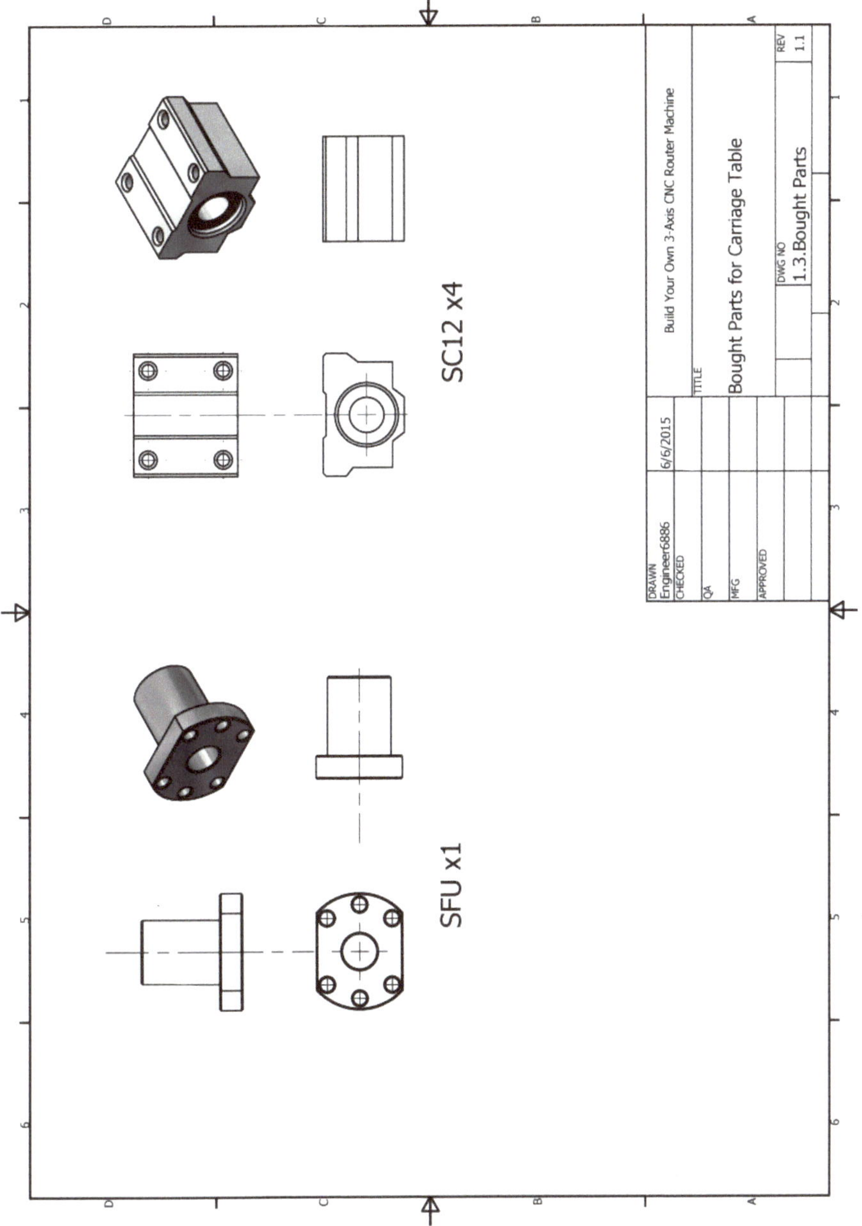
SC12 x4
SFU x1
DRAWN
Engineer6886
6/6/2015
CHECKED
QA
MFG
APPROVED
Build Your Own 3-Axis CNC Router Machine
TITLE
Bought Parts for Carriage Table
DWG NO
1.3.Bought Parts
REV
1.1

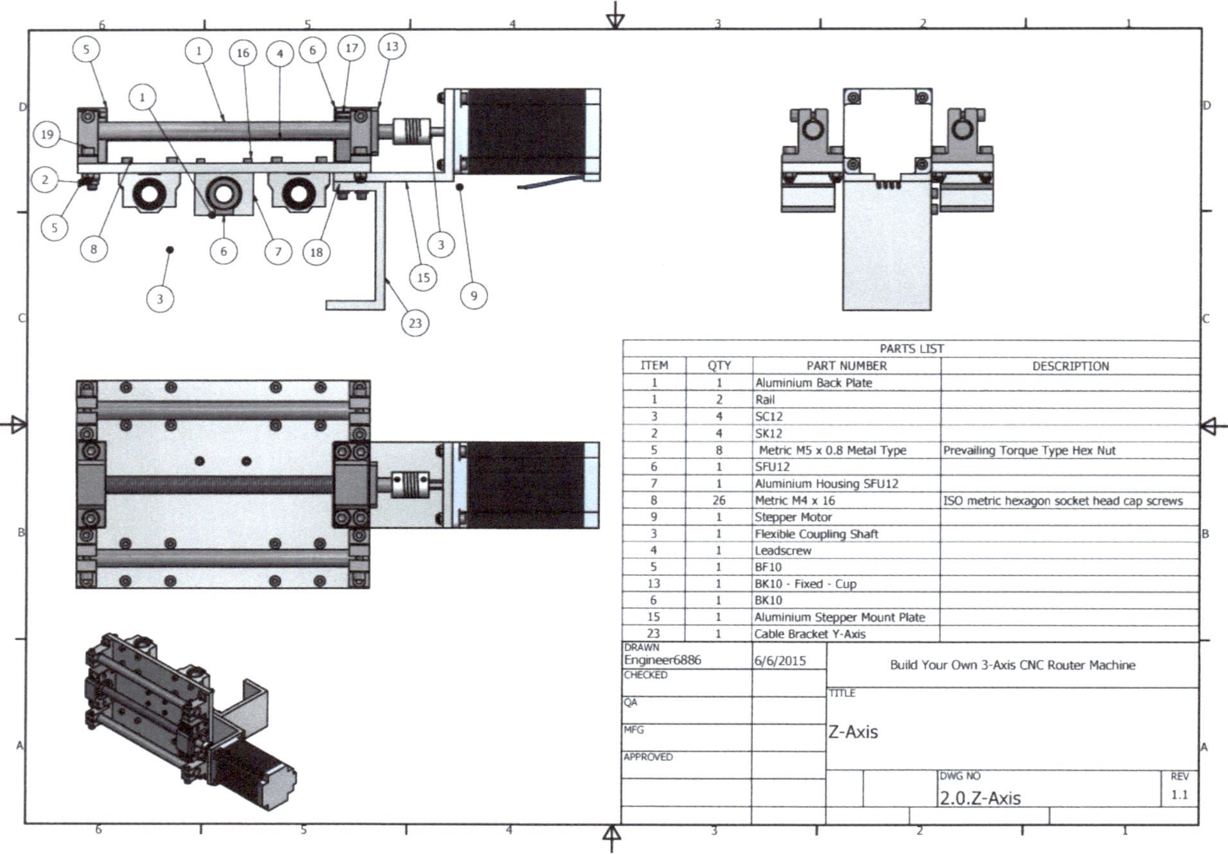

PARTS LIST

ITEM	QTY	PART NUMBER	DESCRIPTION
1	1	Aluminium Back Plate	
1	2	Rail	
3	4	SC12	
2	4	SK12	
5	8	Metric M5 x 0.8 Metal Type	Prevailing Torque Type Hex Nut
6	1	SFU12	
7	1	Aluminium Housing SFU12	
8	26	Metric M4 x 16	ISO metric hexagon socket head cap screws
9	1	Stepper Motor	
3	1	Flexible Coupling Shaft	
4	1	Leadscrew	
5	1	BF10	
13	1	BK10 - Fixed - Cup	
6	1	BK10	
15	1	Aluminium Stepper Mount Plate	
23	1	Cable Bracket Y-Axis	

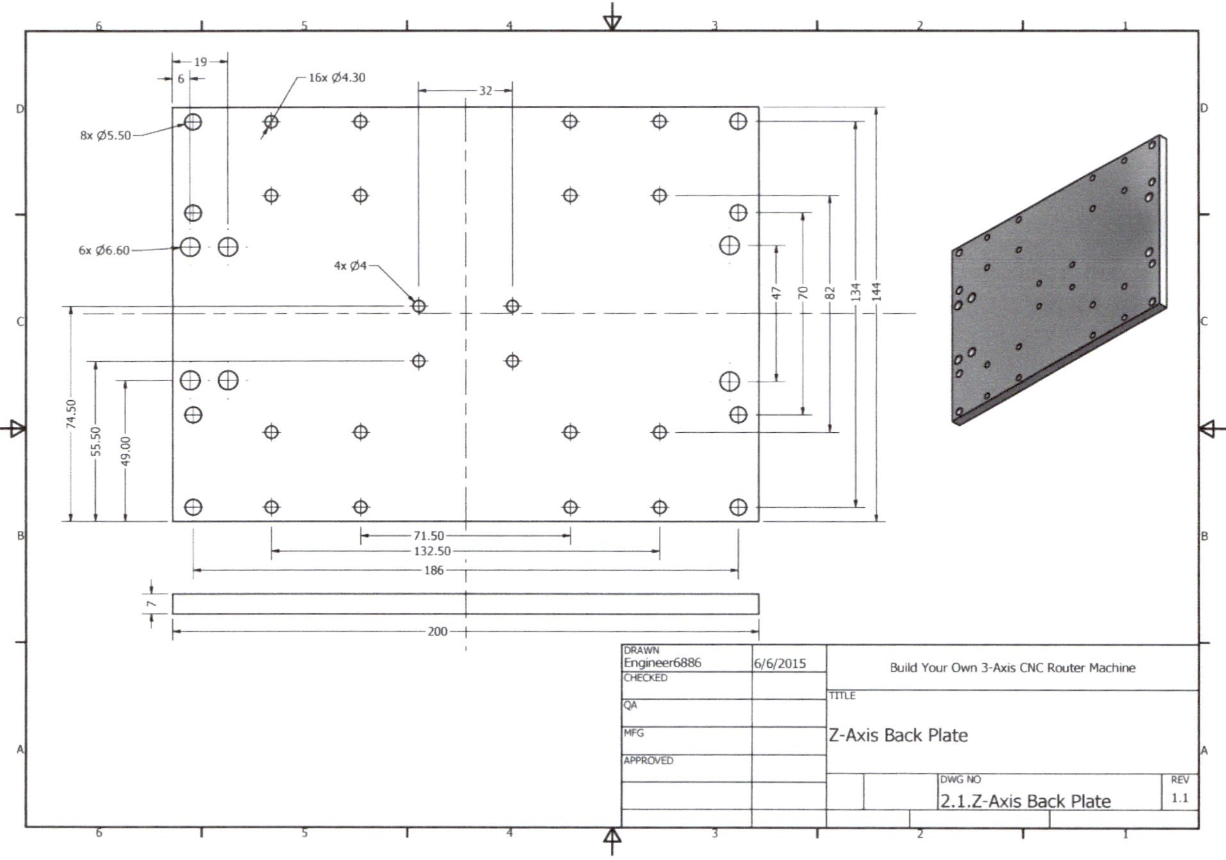
19
6
16x Ø4.30
32
8x Ø5.50
6x Ø6.60
4x Ø4
47
70
82
134
144
74.50
55.50
49.00
71.50
132.50
186
7
200
DRAWN
Engineer6886
6/6/2015
Build Your Own 3-Axis CNC Router Machine
CHECKED
TITLE
QA
MFG
Z-Axis Back Plate
APPROVED
DWG NO
2.1.Z-Axis Back Plate
REV
1.1

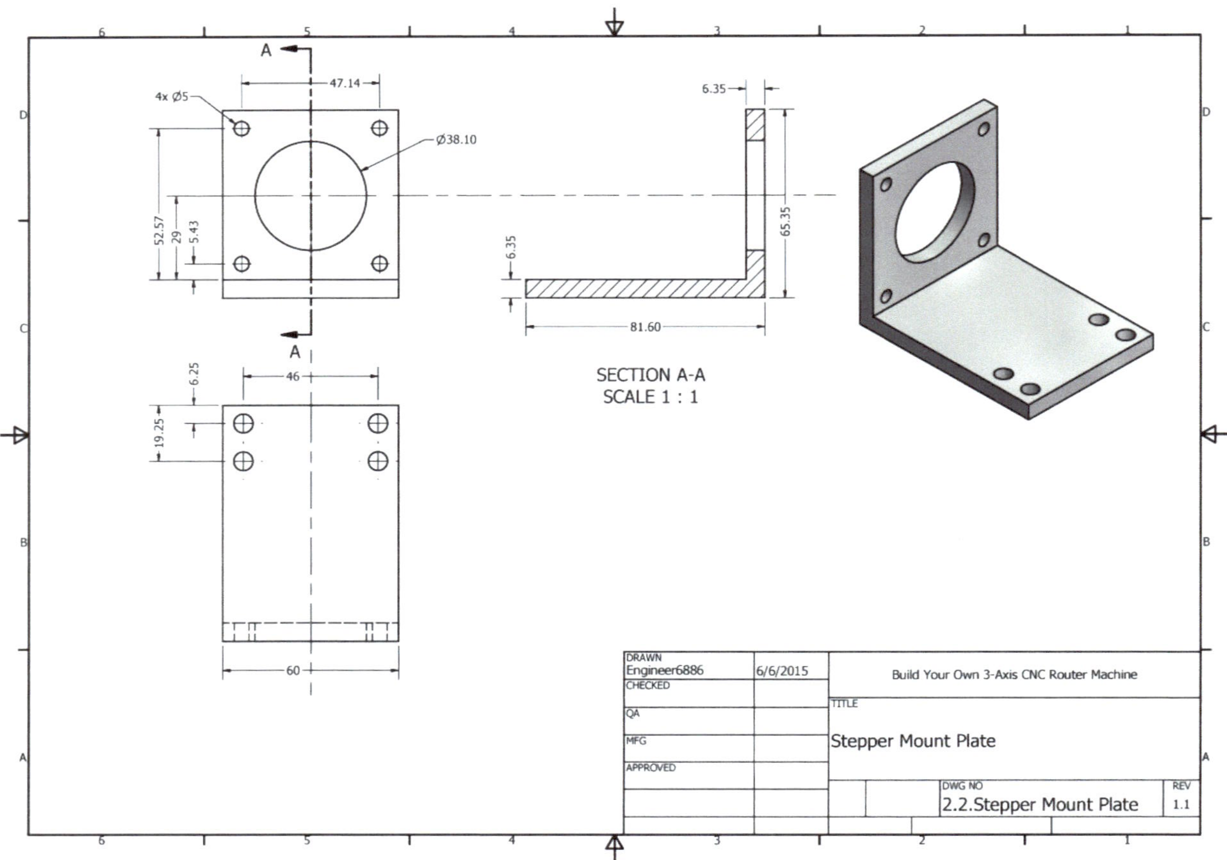

A
47.14
4x Ø5
Ø38.10
52.57
29
5.43
A
6.25
46
19.25
60
6.35
6.35
65.35
81.60
SECTION A-A
SCALE 1 : 1
DRAWN
Engineer6886
6/6/2015
Build Your Own 3-Axis CNC Router Machine
CHECKED
TITLE
QA
MFG
Stepper Mount Plate
APPROVED
DWG NO
2.2.Stepper Mount Plate
REV
1.1

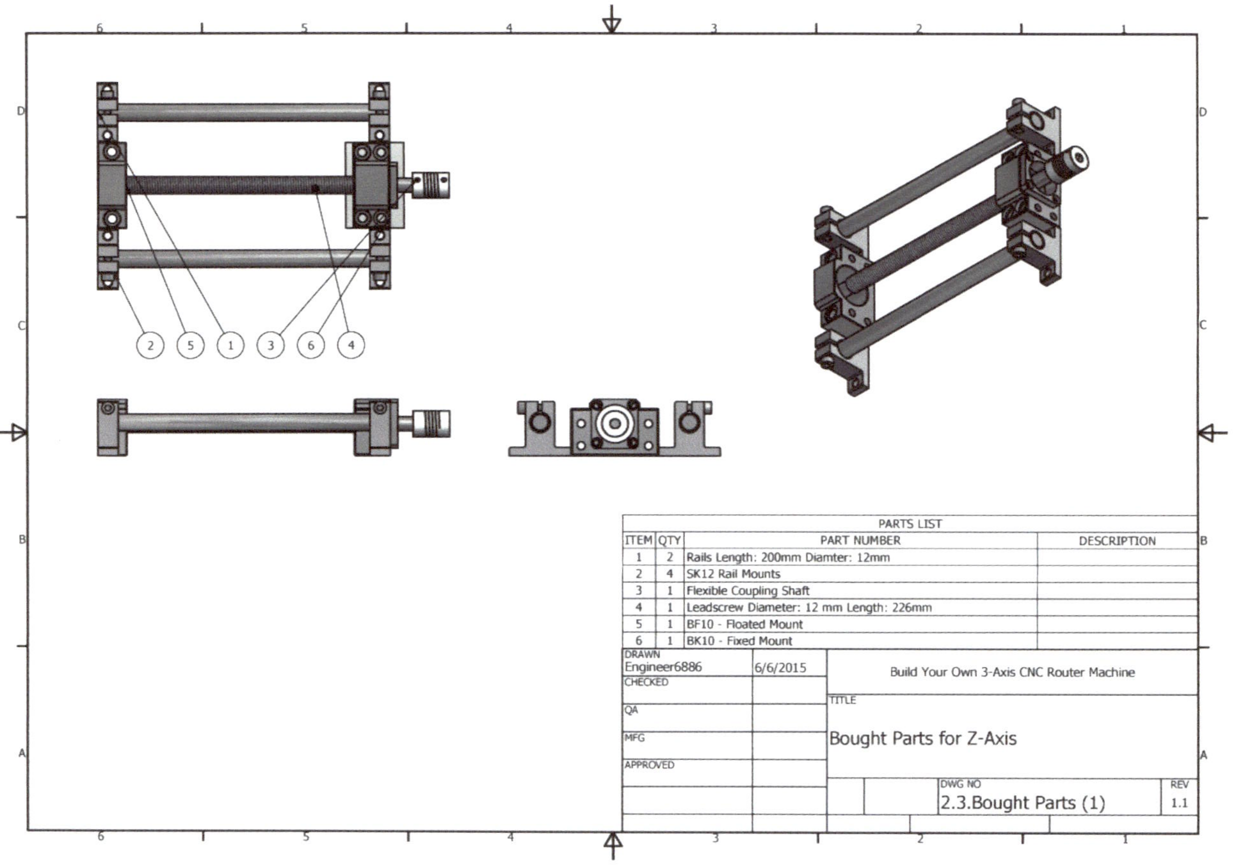
2
5
1
3
6
4
PARTS LIST
ITEM QTY PART NUMBER DESCRIPTION
1 2 Rails Length: 200mm Diamter: 12mm
2 4 SK12 Rail Mounts
3 1 Flexible Coupling Shaft
4 1 Leadscrew Diameter: 12 mm Length: 226mm
5 1 BF10 - Floated Mount
6 1 BK10 - Fixed Mount
DRAWN
Engineer6886
6/6/2015
CHECKED
QA
MFG
APPROVED
Build Your Own 3-Axis CNC Router Machine
TITLE
Bought Parts for Z-Axis
DWG NO
2.3.Bought Parts (1)
REV
1.1

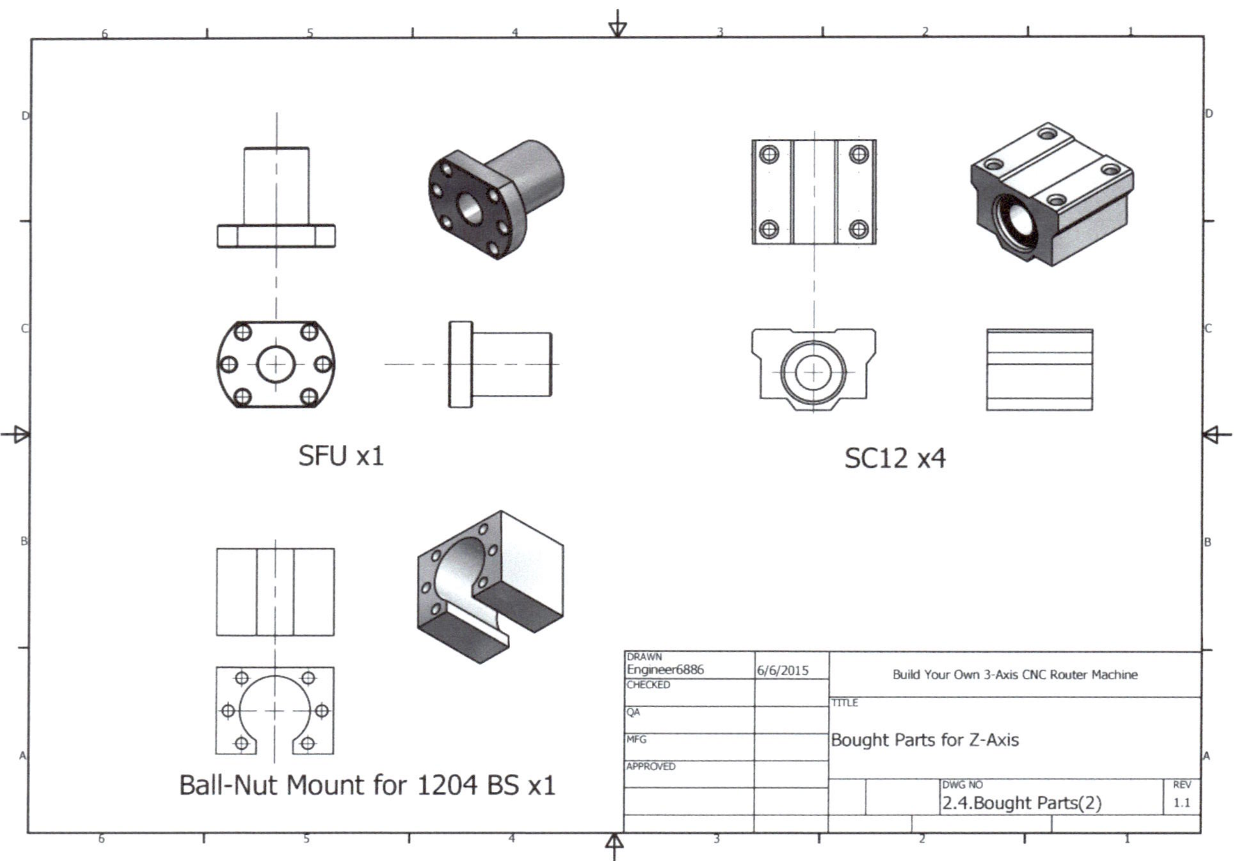
SFU x1
SC12 x4
Ball-Nut Mount for 1204 BS x1
DRAWN
Engineer6886
6/6/2015
CHECKED
QA
MFG
APPROVED
Build Your Own 3-Axis CNC Router Machine
TITLE
Bought Parts for Z-Axis
DWG NO
2.4.Bought Parts(2)
REV
1.1

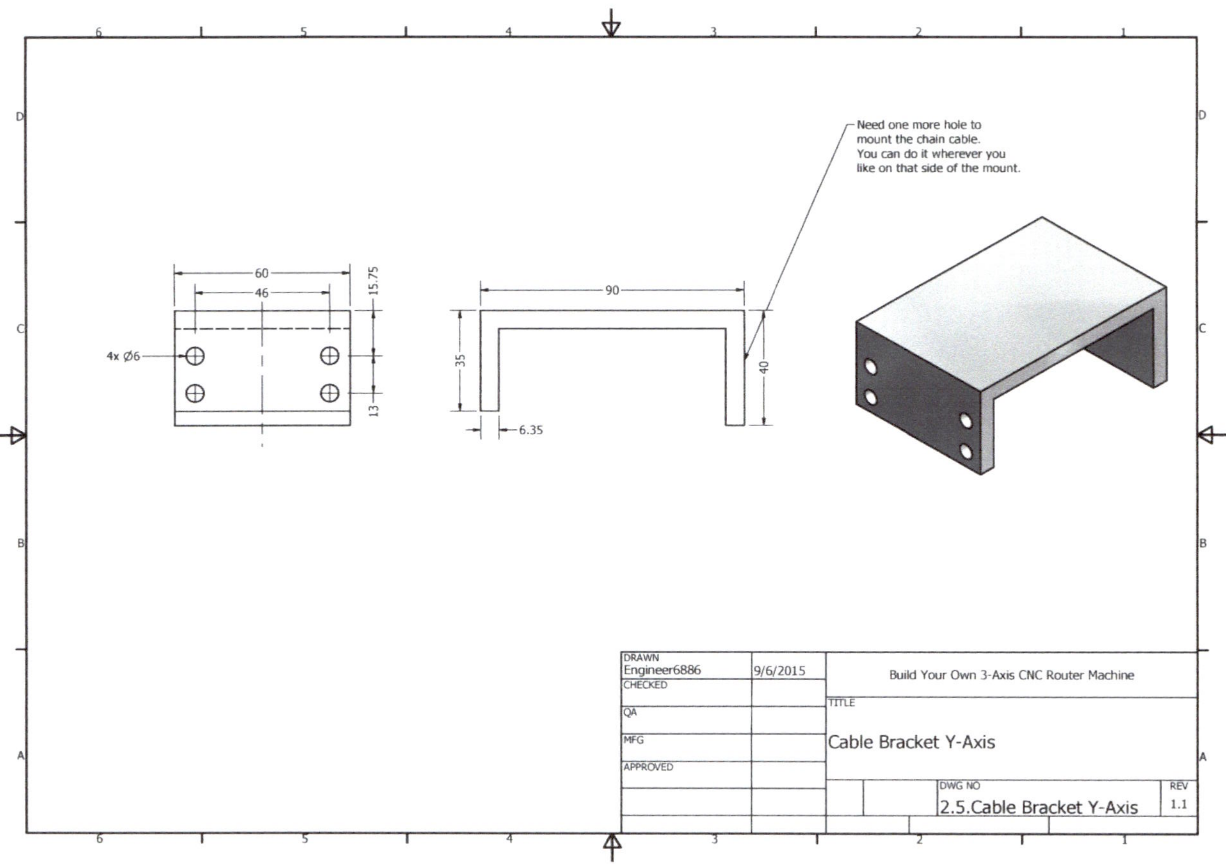
4x Ø6
60
46
15.75
13
90
35
6.35
40
Need one more hole to mount the chain cable. You can do it wherever you like on that side of the mount.
DRAWN
Engineer6886
9/6/2015
CHECKED
QA
MFG
APPROVED
Build Your Own 3-Axis CNC Router Machine
TITLE
Cable Bracket Y-Axis
DWG NO
2.5.Cable Bracket Y-Axis
REV
1.1

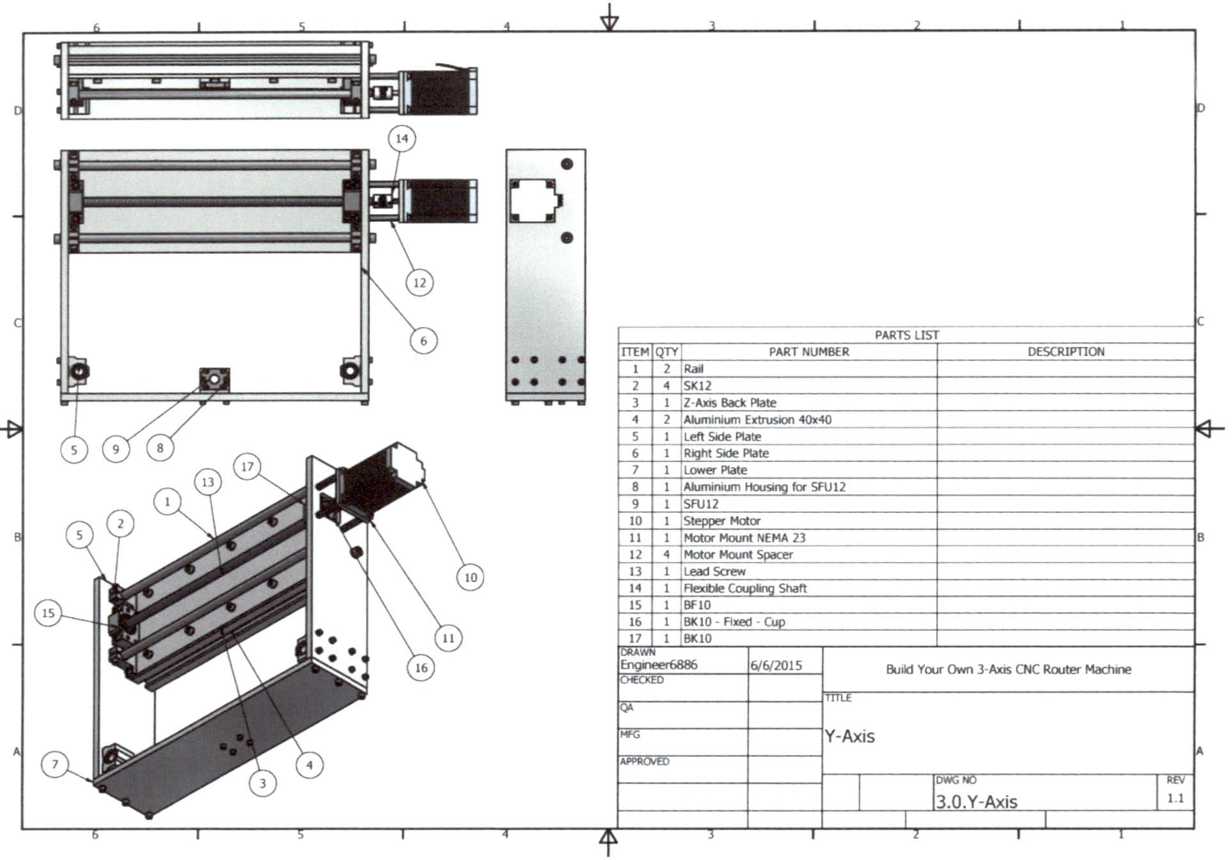

PARTS LIST

ITEM	QTY	PART NUMBER	DESCRIPTION
1	2	Rail	
2	4	SK12	
3	1	Z-Axis Back Plate	
4	2	Aluminium Extrusion 40x40	
5	1	Left Side Plate	
6	1	Right Side Plate	
7	1	Lower Plate	
8	1	Aluminium Housing for SFU12	
9	1	SFU12	
10	1	Stepper Motor	
11	1	Motor Mount NEMA 23	
12	4	Motor Mount Spacer	
13	1	Lead Screw	
14	1	Flexible Coupling Shaft	
15	1	BF10	
16	1	BK10 - Fixed - Cup	
17	1	BK10	

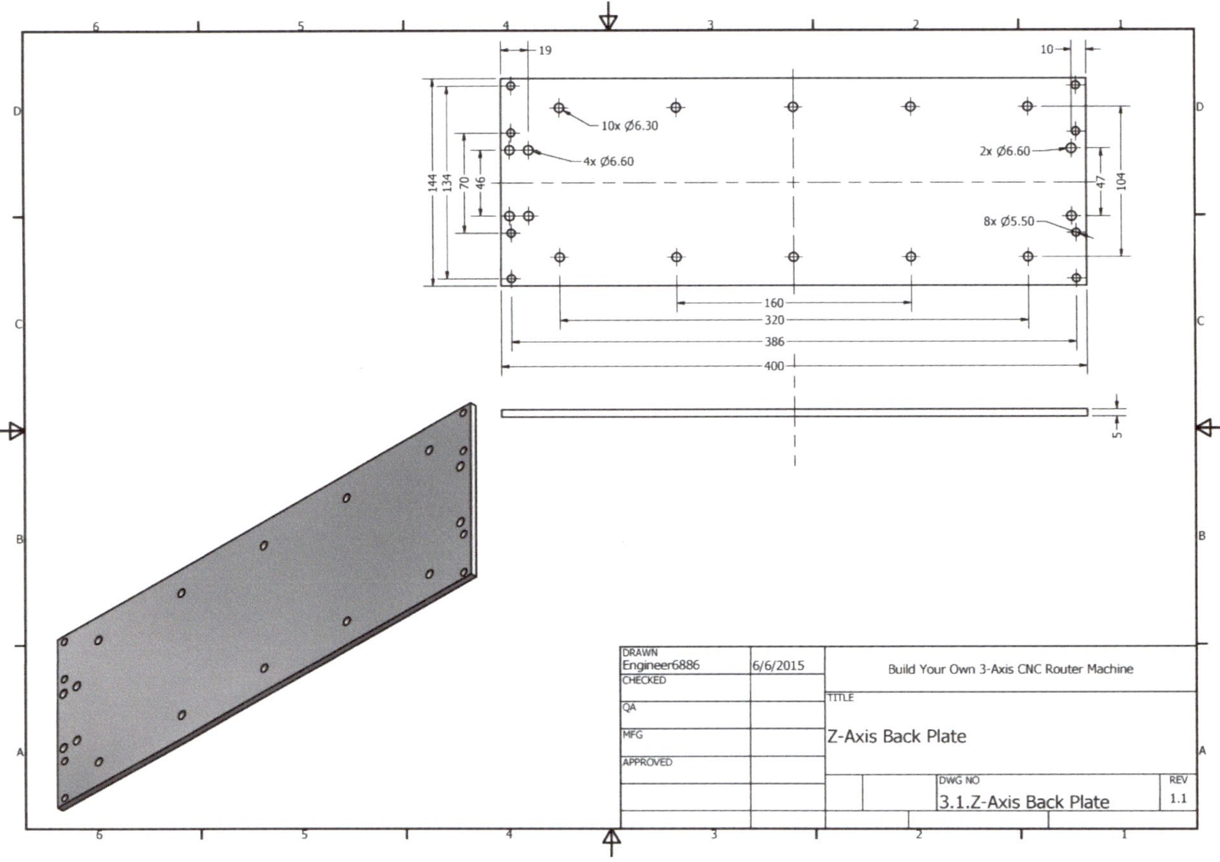

19
10
10x Ø6.30
4x Ø6.60
2x Ø6.60
8x Ø5.50
144
134
70
46
47
104
160
320
386
400
5
DRAWN
Engineer6886
6/6/2015
CHECKED
QA
MFG
APPROVED
Build Your Own 3-Axis CNC Router Machine
TITLE
Z-Axis Back Plate
DWG NO
3.1.Z-Axis Back Plate
REV
1.1

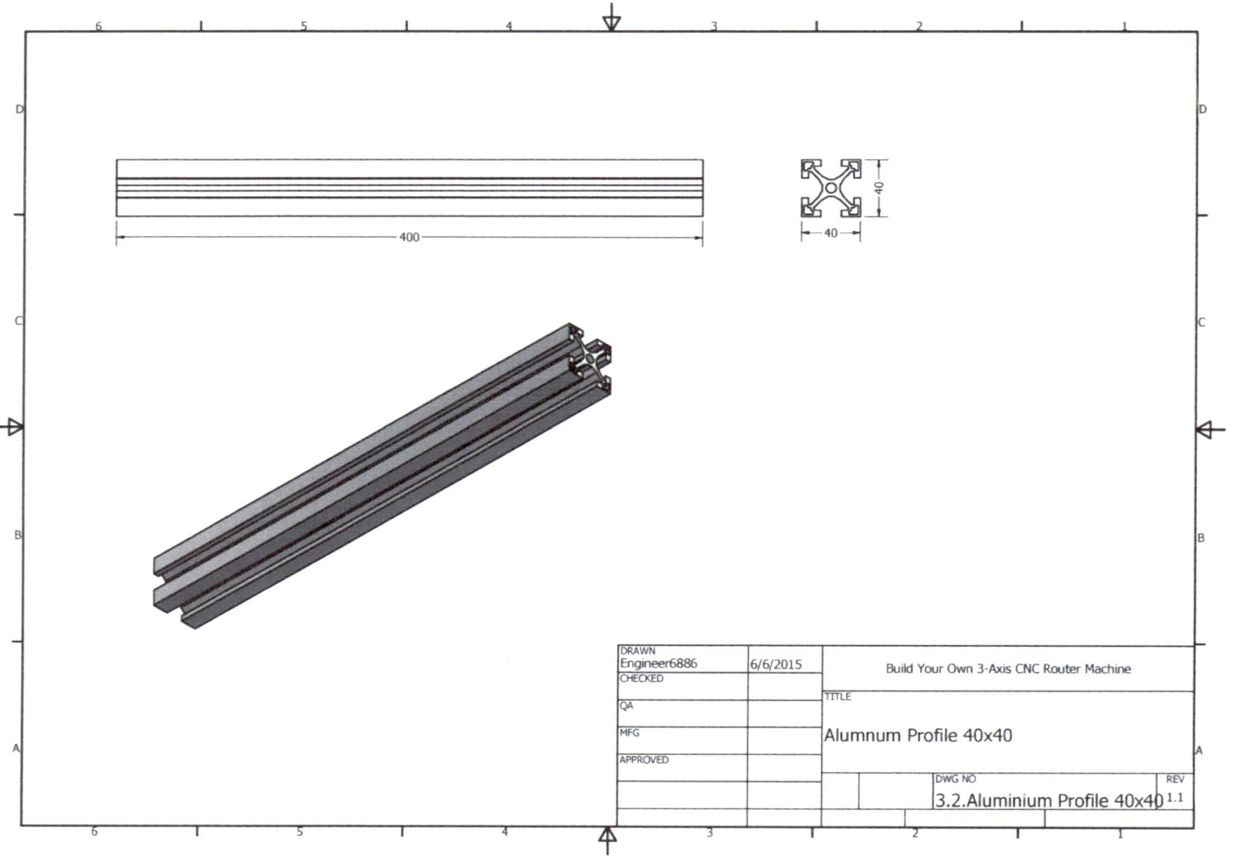

400
40
40
DRAWN
Engineer6886
6/6/2015
Build Your Own 3-Axis CNC Router Machine
CHECKED
TITLE
QA
MFG
Alumnum Profile 40x40
APPROVED
DWG NO
3.2.Aluminium Profile 40x40
REV
1.1

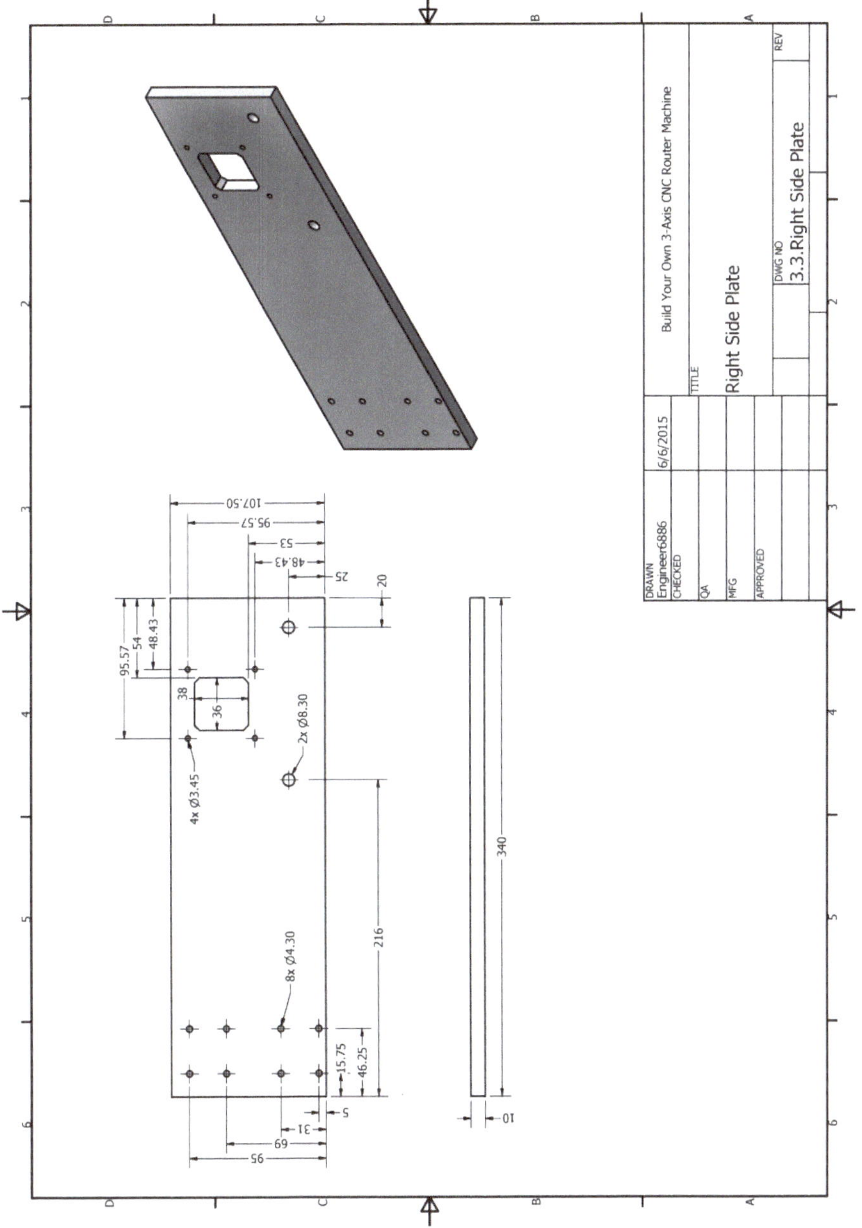
Build Your Own 3-Axis CNC Router Machine
DRAWN
Engineer6886
6/6/2015
CHECKED
QA
MFG
APPROVED
TITLE
Right Side Plate
DWG NO
3.3.Right Side Plate
REV
4x Ø3.45
2x Ø8.30
8x Ø4.30
107.50
95.57
53
48.43
25
20
54
38
36
216
340
15.75
46.25
5
31
69
95
10

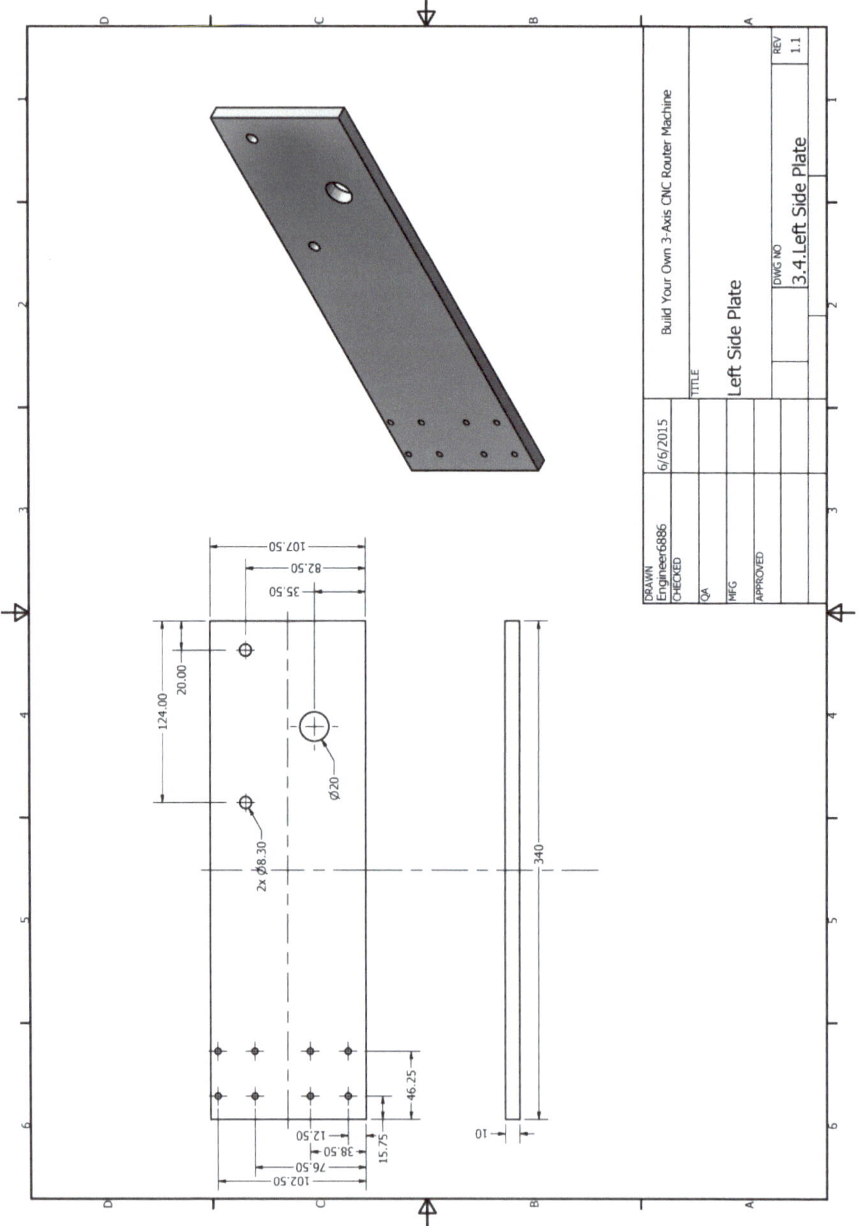
Build Your Own 3-Axis CNC Router Machine
TITLE
Left Side Plate
DWG NO
3.4.Left Side Plate
REV
1.1
DRAWN
Engineer6886
6/6/2015
CHECKED
QA
MFG
APPROVED
107.50
82.50
35.50
124.00
20.00
Ø20
2x Ø8.30
340
46.25
15.75
12.50
38.50
76.50
102.50
10

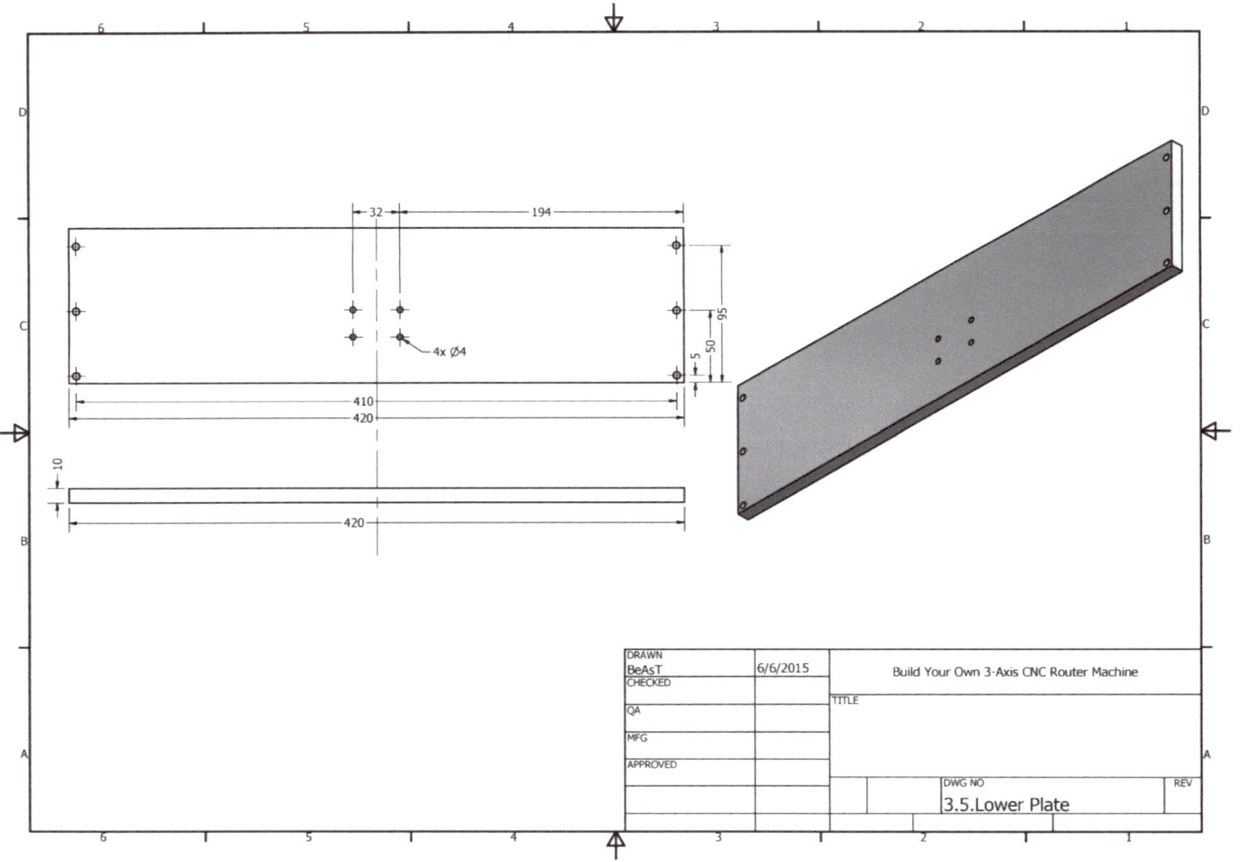

32
194
4x Ø4
95
50
5
410
420
10
420
DRAWN
BeAsT
6/6/2015
Build Your Own 3-Axis CNC Router Machine
CHECKED
TITLE
QA
MFG
APPROVED
DWG NO
3.5.Lower Plate
REV

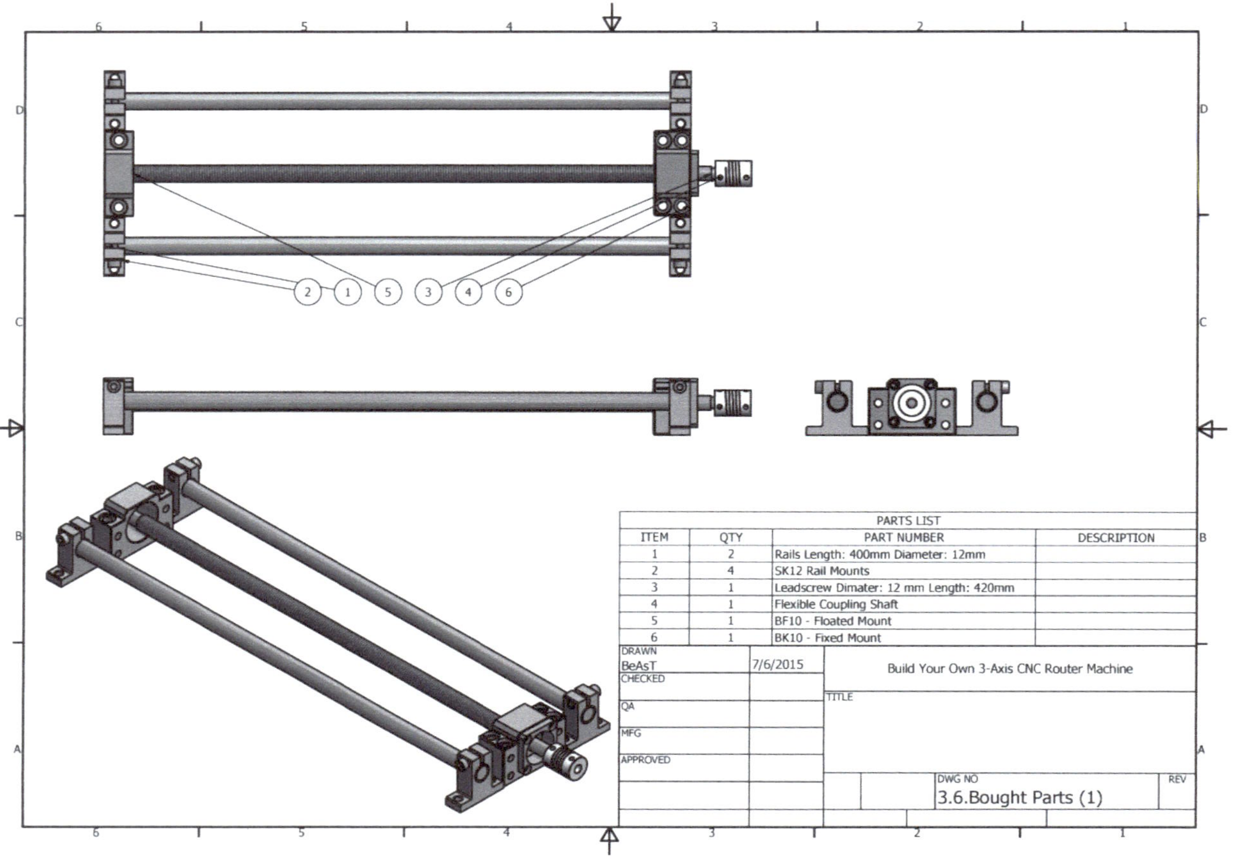
2
1
5
3
4
6
PARTS LIST
ITEM | QTY | PART NUMBER | DESCRIPTION
1 | 2 | Rails Length: 400mm Diameter: 12mm
2 | 4 | SK12 Rail Mounts
3 | 1 | Leadscrew Dimater: 12 mm Length: 420mm
4 | 1 | Flexible Coupling Shaft
5 | 1 | BF10 - Floated Mount
6 | 1 | BK10 - Fixed Mount
DRAWN
BeAsT
7/6/2015
Build Your Own 3-Axis CNC Router Machine
CHECKED
TITLE
QA
MFG
APPROVED
DWG NO
3.6.Bought Parts (1)
REV

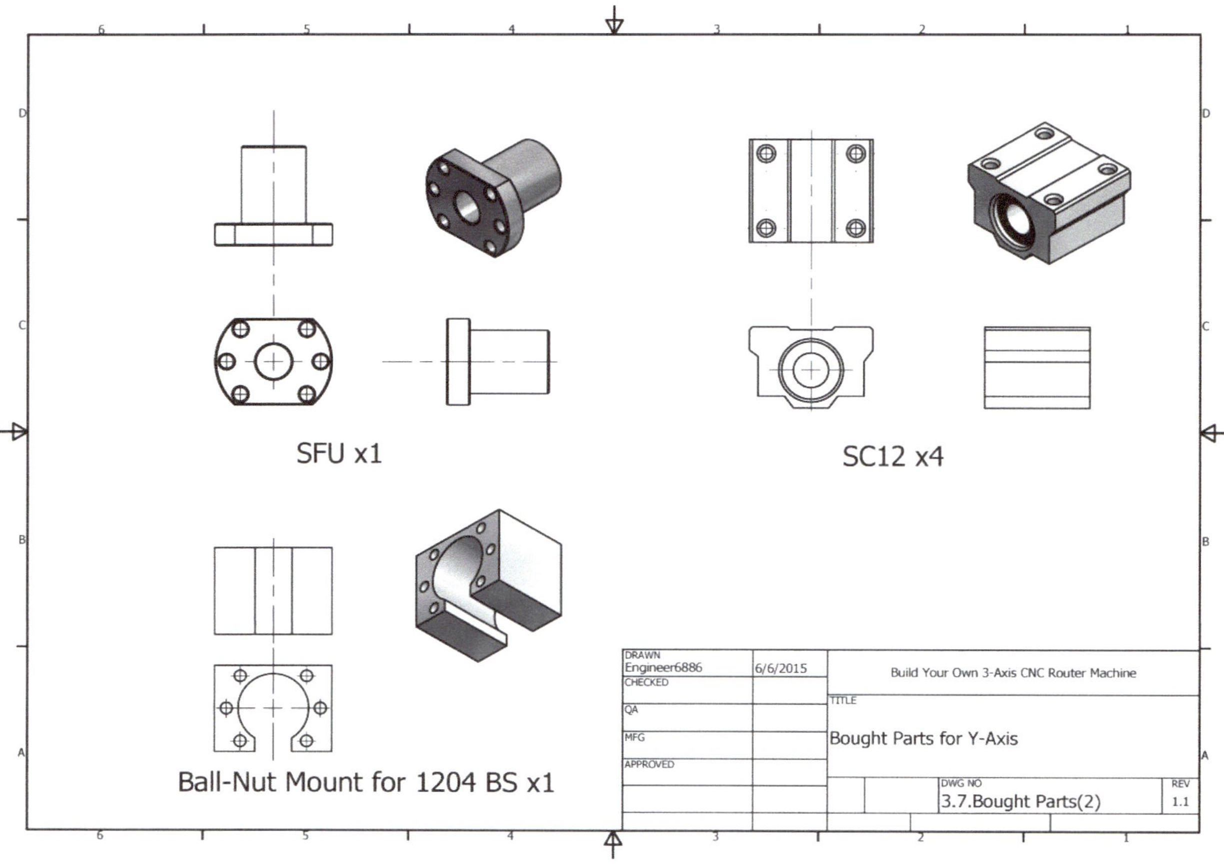
SFU x1
SC12 x4
Ball-Nut Mount for 1204 BS x1
DRAWN
Engineer6886
6/6/2015
CHECKED
QA
MFG
APPROVED
Build Your Own 3-Axis CNC Router Machine
TITLE
Bought Parts for Y-Axis
DWG NO
3.7.Bought Parts(2)
REV
1.1

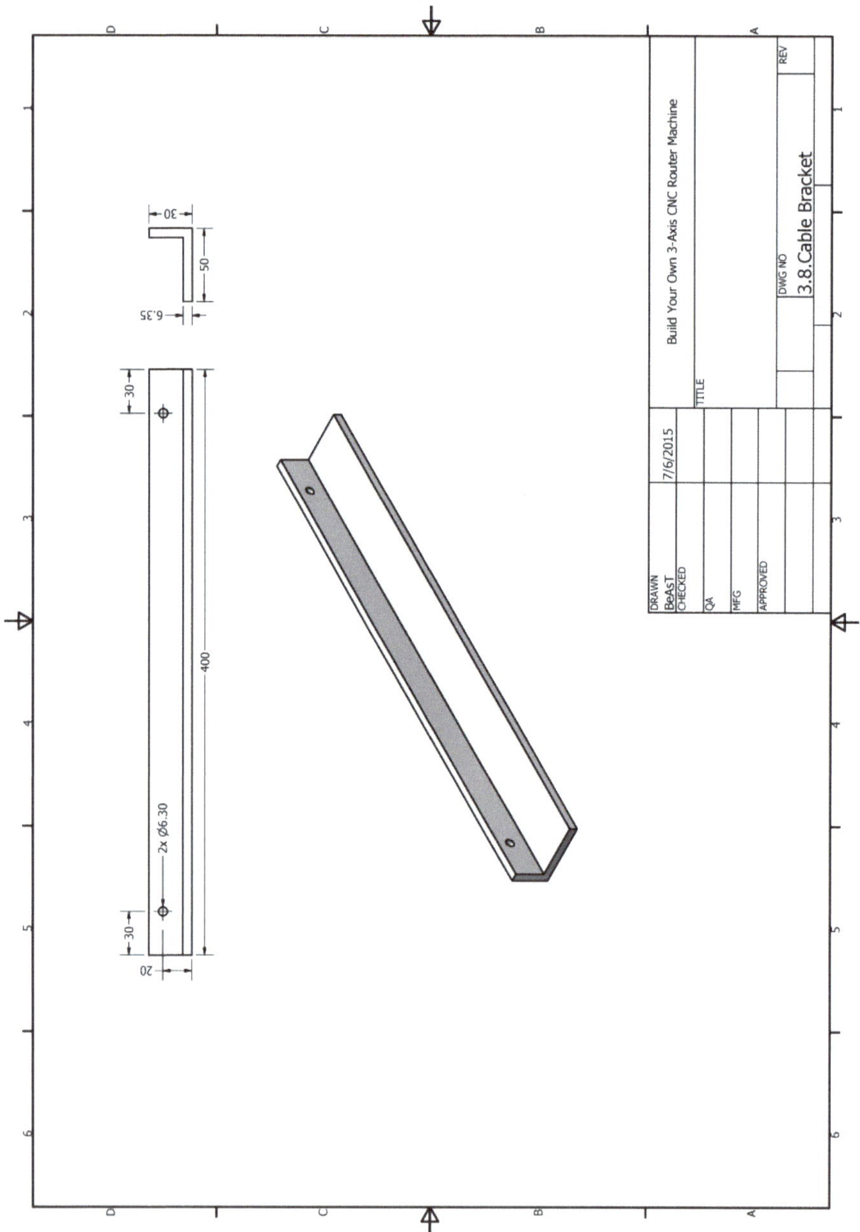
30
2x Ø6.30
30
20
400
30
50
6.35
DRAWN
BeAsT
7/6/2015
CHECKED
QA
MFG
APPROVED
TITLE
Build Your Own 3-Axis CNC Router Machine
DWG NO
3.8.Cable Bracket
REV

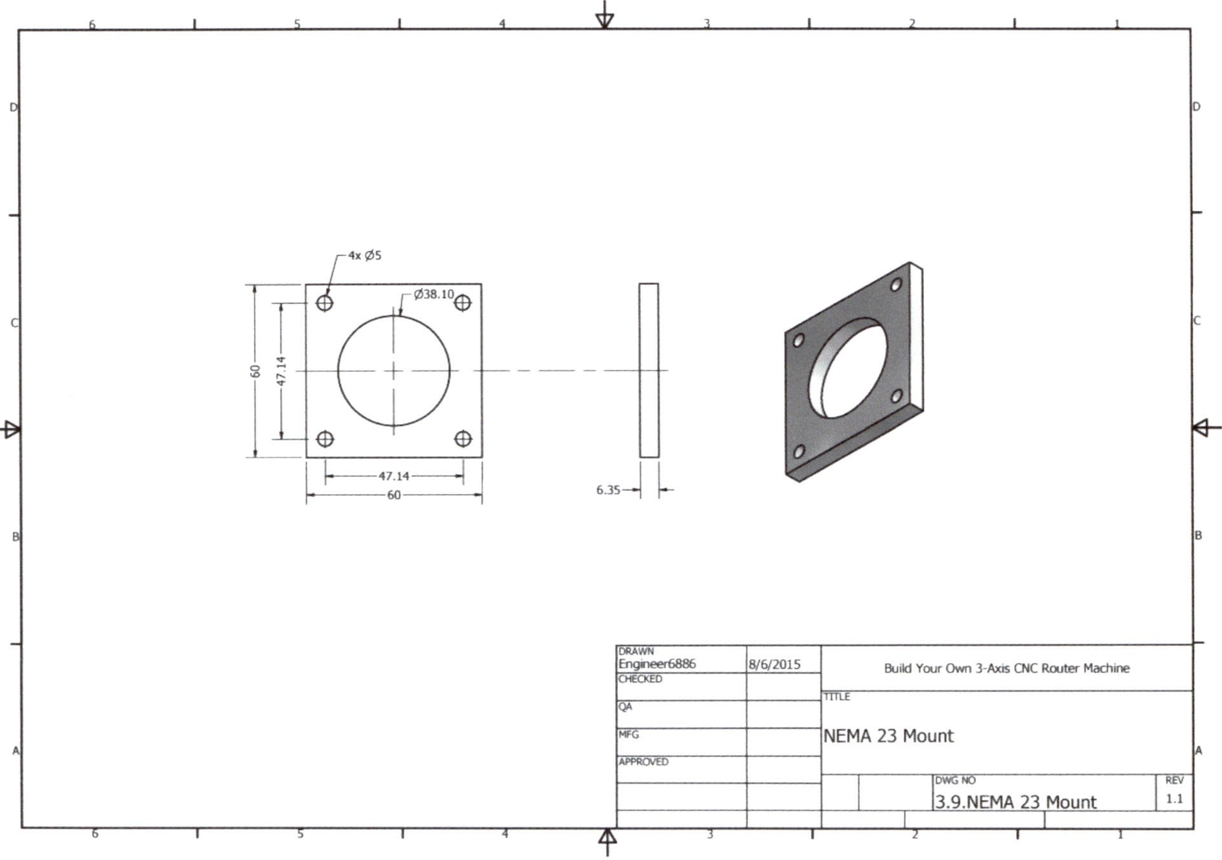

4x Ø5
Ø38.10
60
47.14
47.14
60
6.35
DRAWN
Engineer6886
8/6/2015
CHECKED
QA
MFG
APPROVED
Build Your Own 3-Axis CNC Router Machine
TITLE
NEMA 23 Mount
DWG NO
3.9.NEMA 23 Mount
REV
1.1

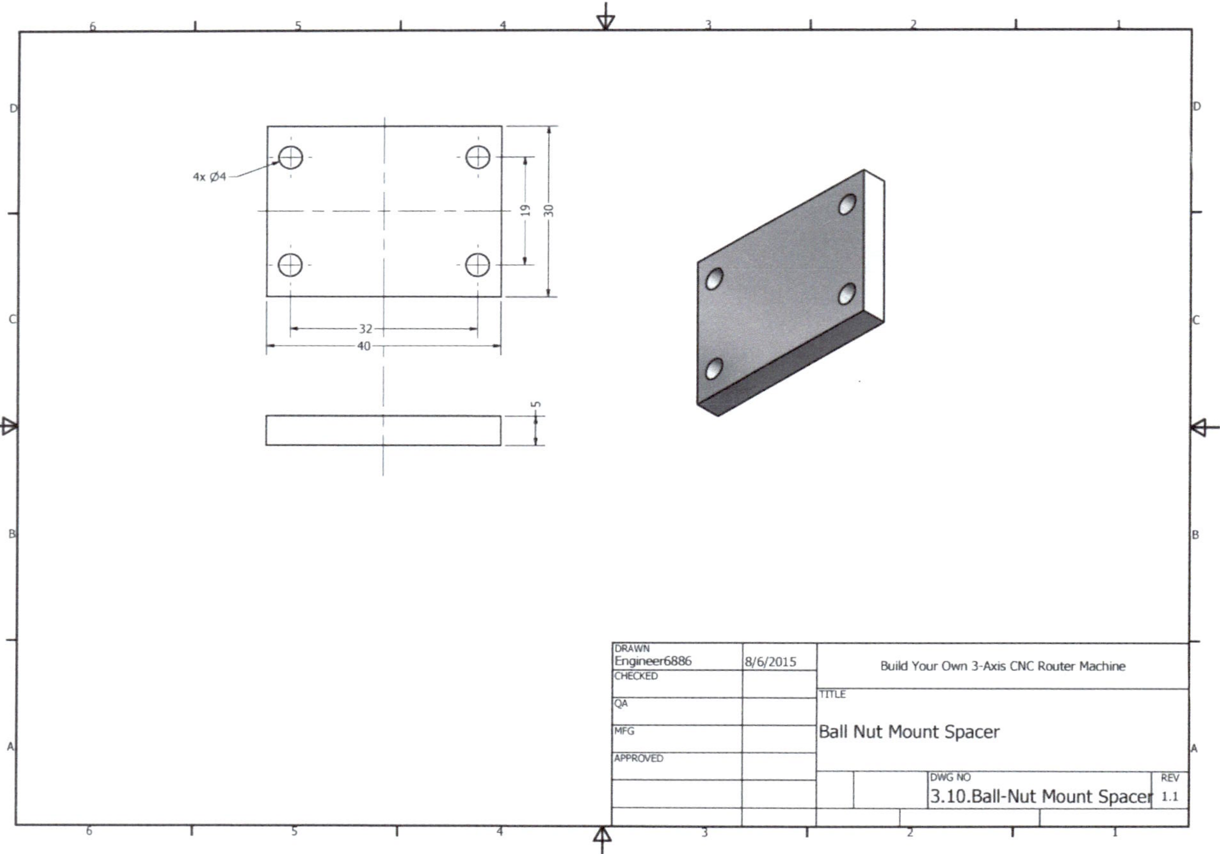
4x Ø4
19
30
32
40
5
DRAWN
Engineer6886
8/6/2015
CHECKED
QA
MFG
APPROVED
Build Your Own 3-Axis CNC Router Machine
TITLE
Ball Nut Mount Spacer
DWG NO
3.10.Ball-Nut Mount Spacer
REV
1.1

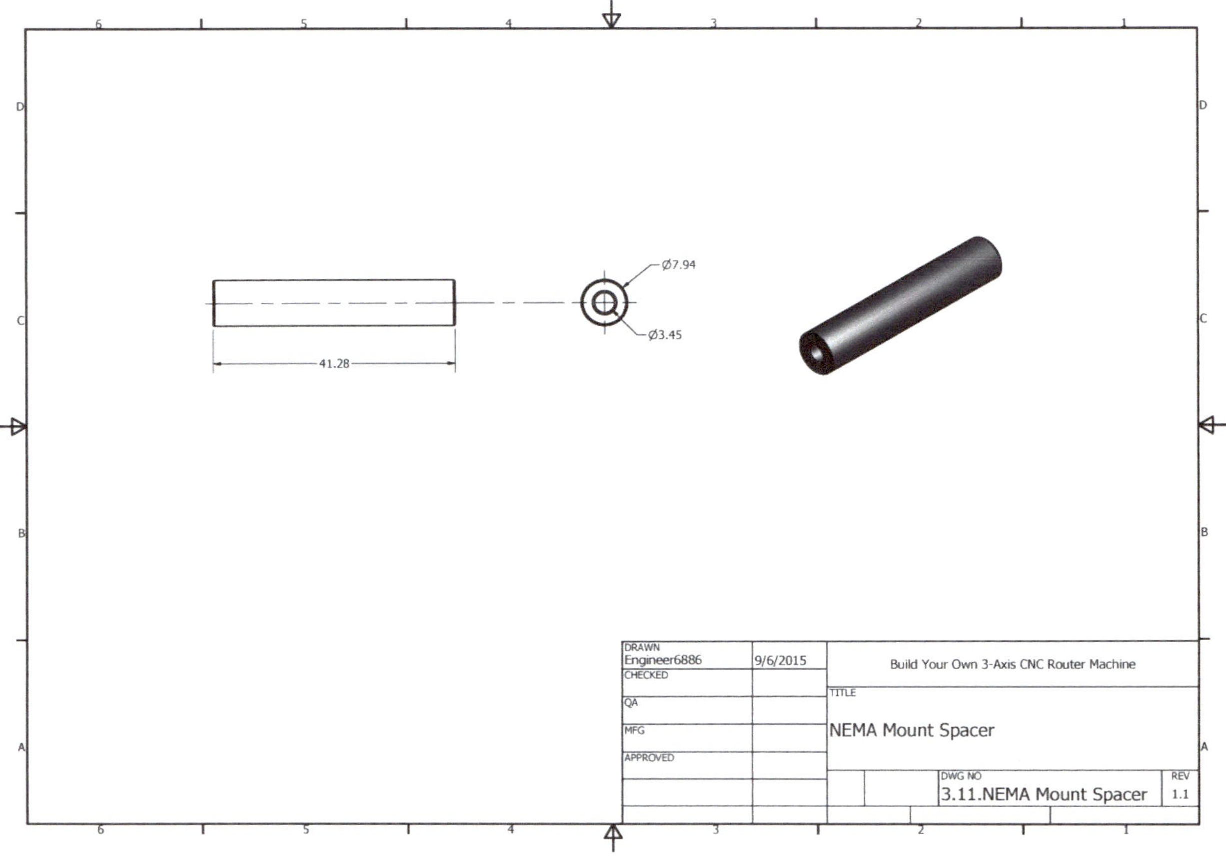
Ø7.94
Ø3.45
41.28
DRAWN
Engineer6886
9/6/2015
CHECKED
QA
MFG
APPROVED
Build Your Own 3-Axis CNC Router Machine
TITLE
NEMA Mount Spacer
DWG NO
3.11.NEMA Mount Spacer
REV
1.1

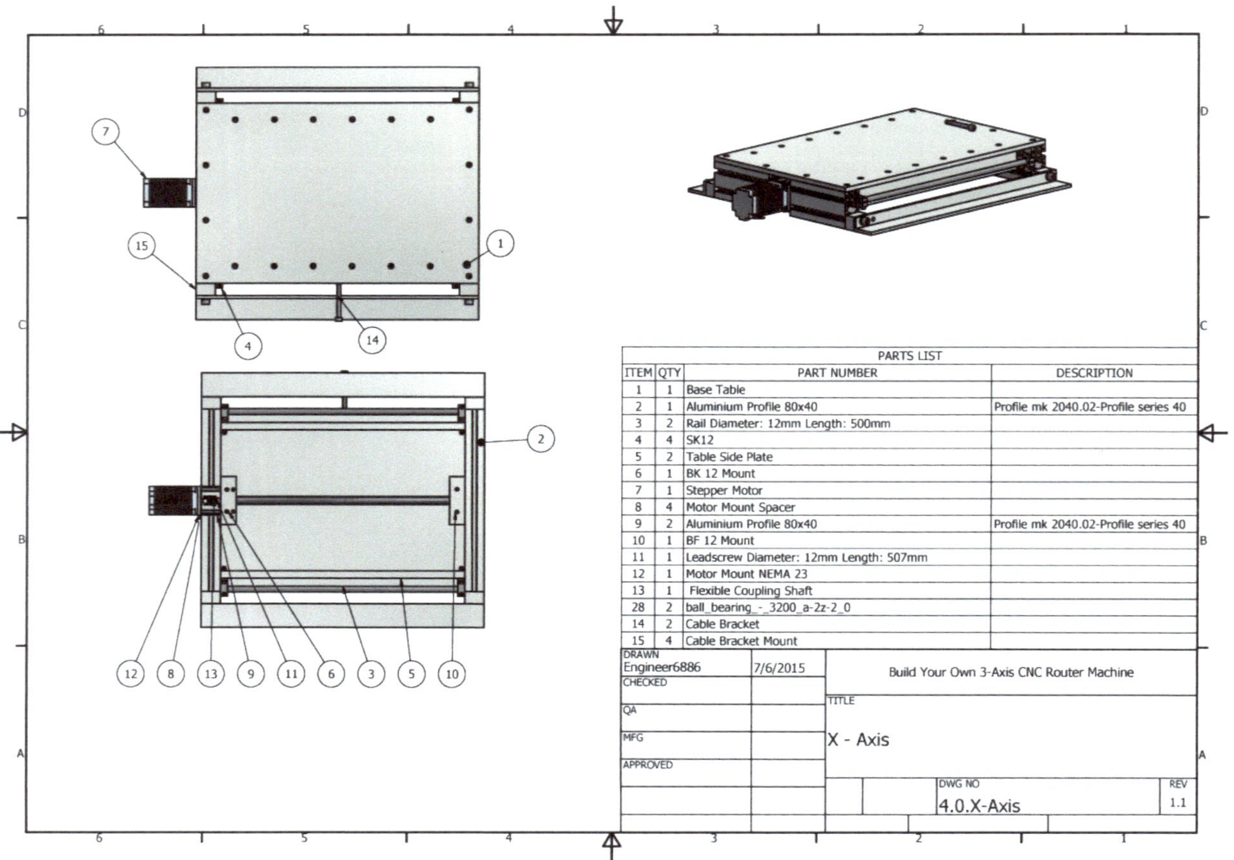

PARTS LIST

ITEM	QTY	PART NUMBER	DESCRIPTION
1	1	Base Table	
2	1	Aluminium Profile 80x40	Profile mk 2040.02-Profile series 40
3	2	Rail Diameter: 12mm Length: 500mm	
4	4	SK12	
5	2	Table Side Plate	
6	1	BK 12 Mount	
7	1	Stepper Motor	
8	4	Motor Mount Spacer	
9	2	Aluminium Profile 80x40	Profile mk 2040.02-Profile series 40
10	1	BF 12 Mount	
11	1	Leadscrew Diameter: 12mm Length: 507mm	
12	1	Motor Mount NEMA 23	
13	1	Flexible Coupling Shaft	
28	2	ball_bearing_-_3200_a-2z-2_0	
14	2	Cable Bracket	
15	4	Cable Bracket Mount	

DRAWN		
Engineer6886	7/6/2015	Build Your Own 3-Axis CNC Router Machine
CHECKED		TITLE
QA		X - Axis
MFG		
APPROVED		DWG NO 4.0.X-Axis — REV 1.1

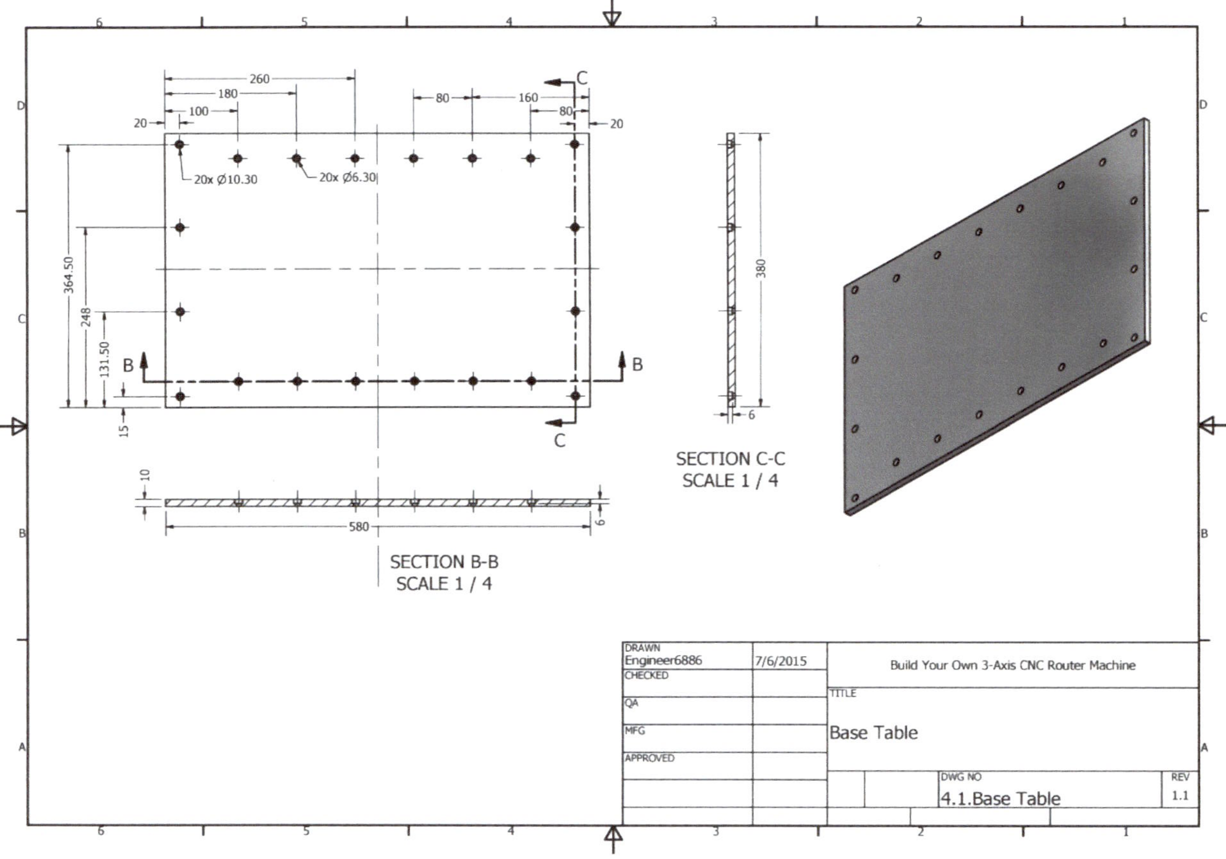

260
180
100
20
80
160
80
20
20x Ø10.30
20x Ø6.30
364.50
248
131.50
15
10
580
6
SECTION B-B
SCALE 1 / 4
380
6
SECTION C-C
SCALE 1 / 4
DRAWN
Engineer6886
7/6/2015
CHECKED
QA
MFG
APPROVED
Build Your Own 3-Axis CNC Router Machine
TITLE
Base Table
DWG NO
4.1.Base Table
REV
1.1

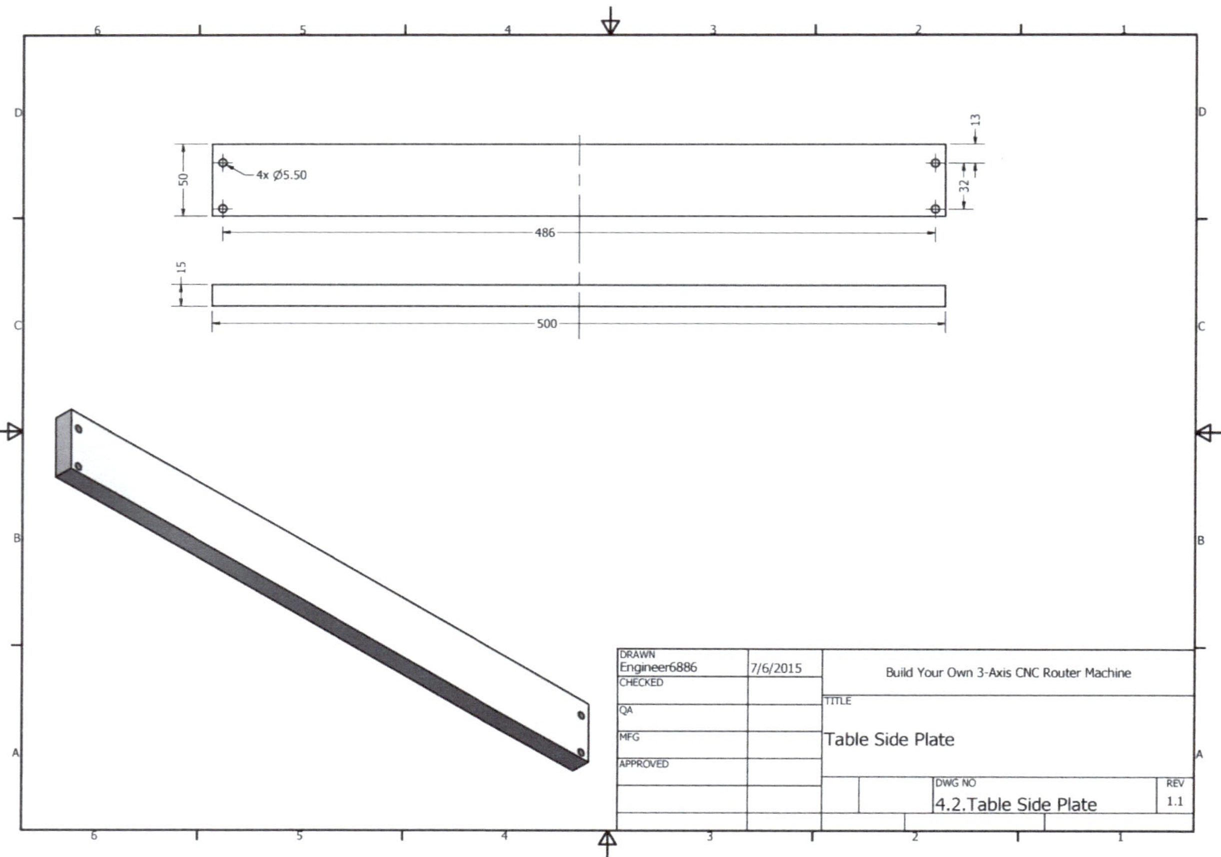
4x Ø5.50
50
486
13
32
15
500
DRAWN
Engineer6886
7/6/2015
CHECKED
QA
MFG
APPROVED
Build Your Own 3-Axis CNC Router Machine
TITLE
Table Side Plate
DWG NO
4.2.Table Side Plate
REV
1.1

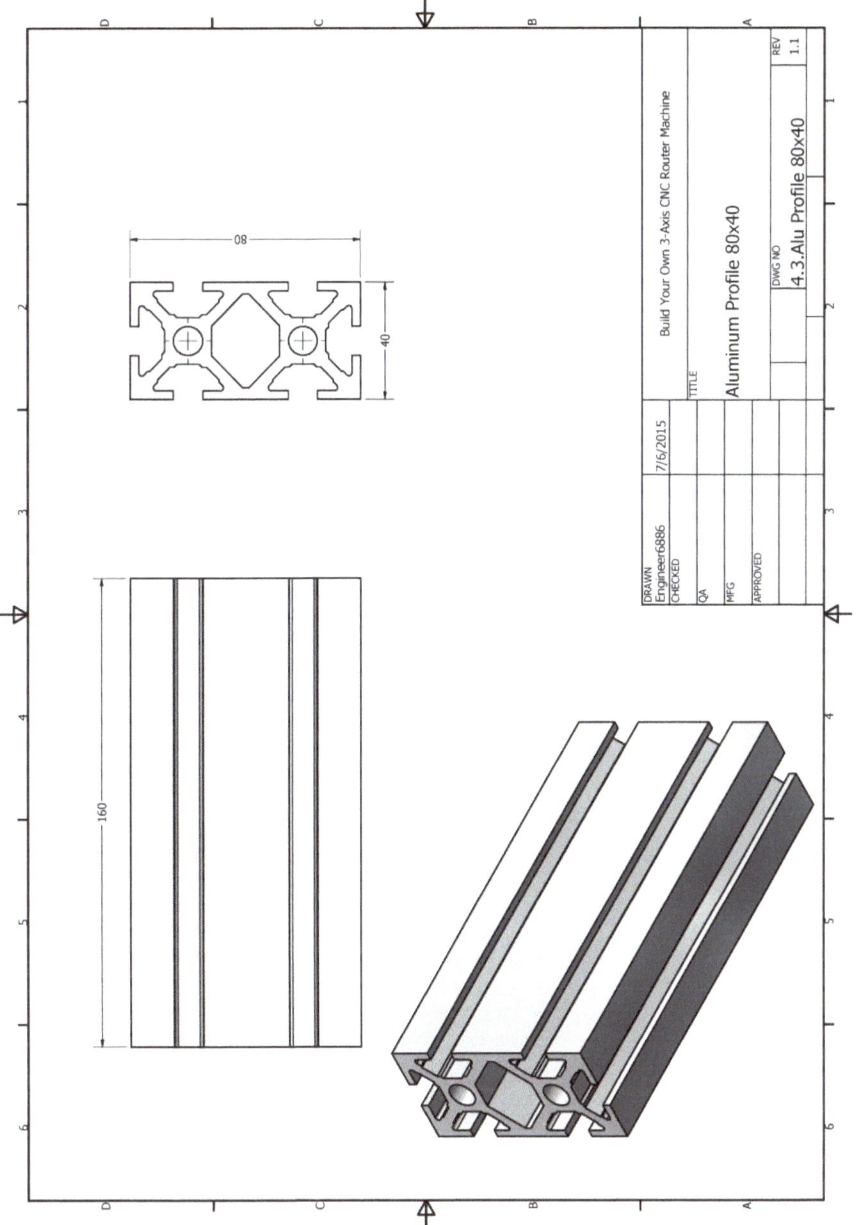
80
40
160
DRAWN
Engineer6886
7/6/2015
CHECKED
QA
MFG
APPROVED
Build Your Own 3-Axis CNC Router Machine
TITLE
Aluminum Profile 80x40
DWG NO
4.3.Alu Profile 80x40
REV
1.1

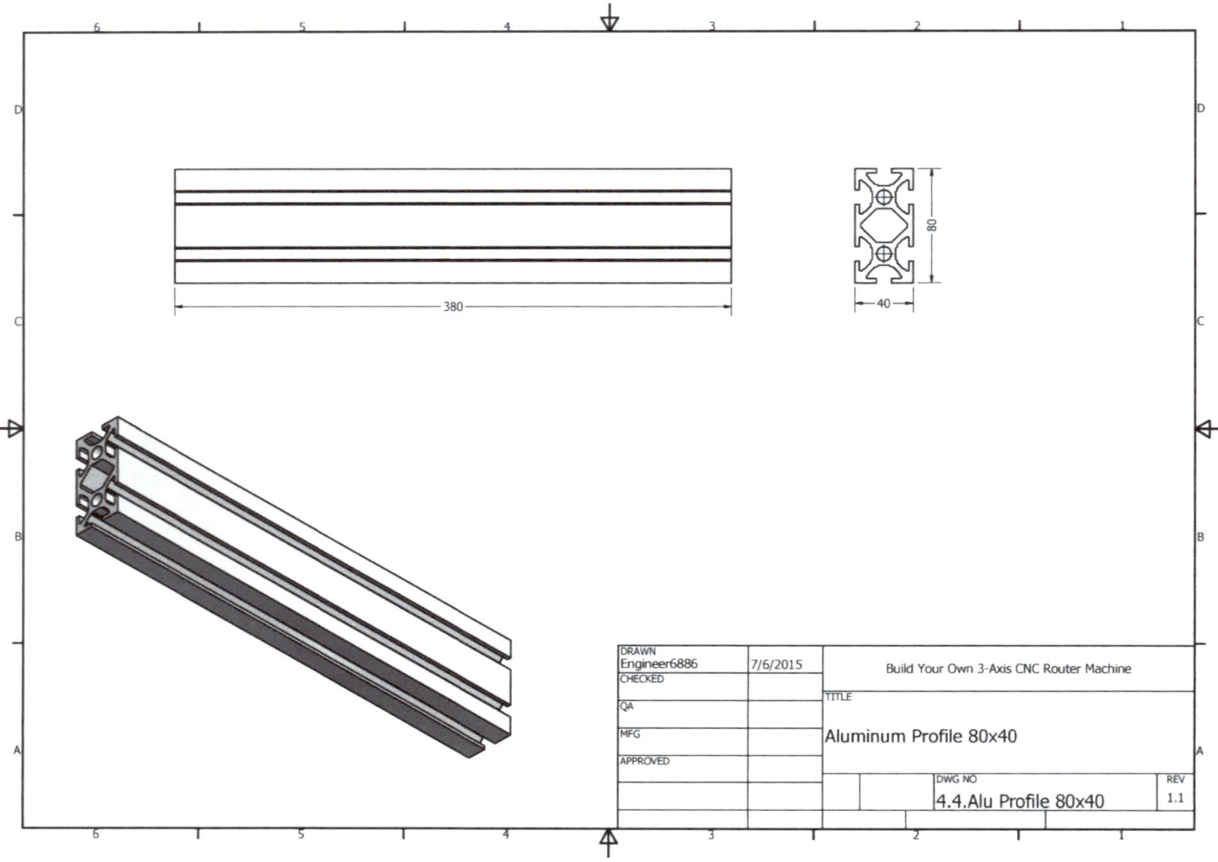

380
80
40
DRAWN
Engineer6886
7/6/2015
CHECKED
QA
MFG
APPROVED
Build Your Own 3-Axis CNC Router Machine
TITLE
Aluminum Profile 80x40
DWG NO
4.4.Alu Profile 80x40
REV
1.1

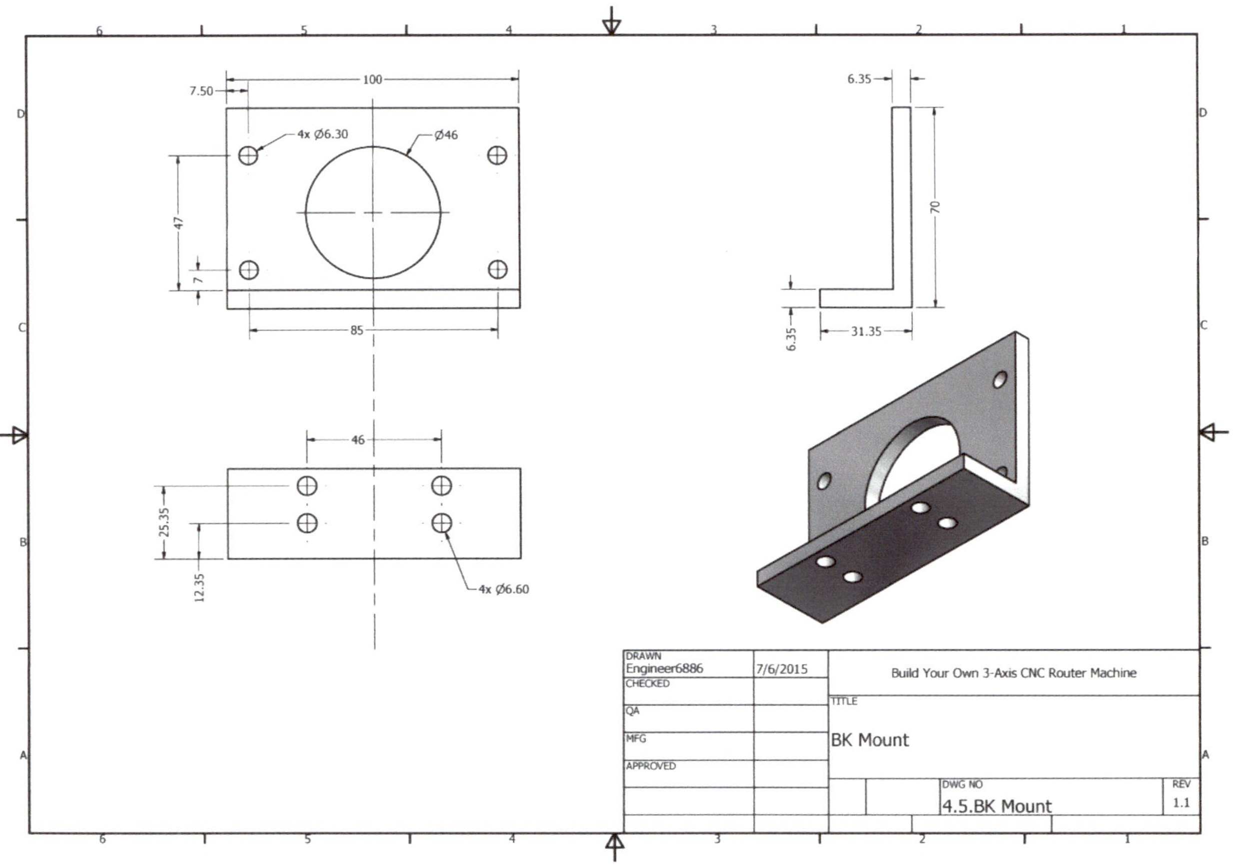
100
7.50
4x Ø6.30
Ø46
47
7
85
46
25.35
12.35
4x Ø6.60
6.35
70
6.35
31.35
DRAWN
Engineer6886
7/6/2015
CHECKED
QA
MFG
APPROVED
Build Your Own 3-Axis CNC Router Machine
TITLE
BK Mount
DWG NO
4.5.BK Mount
REV
1.1

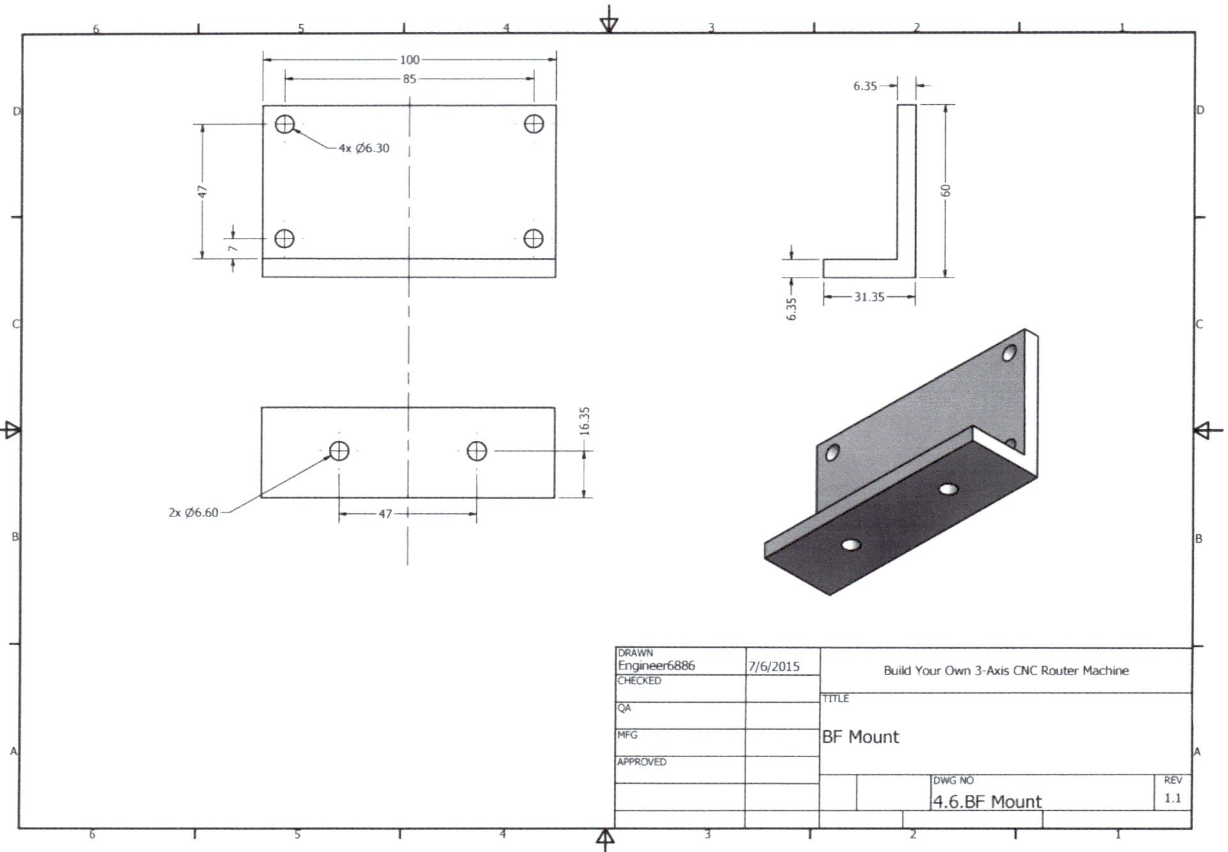
100
85
4x Ø6.30
47
7
6.35
60
6.35
31.35
16.35
2x Ø6.60
47
DRAWN
Engineer6886
7/6/2015
CHECKED
QA
MFG
APPROVED
Build Your Own 3-Axis CNC Router Machine
TITLE
BF Mount
DWG NO
4.6.BF Mount
REV
1.1

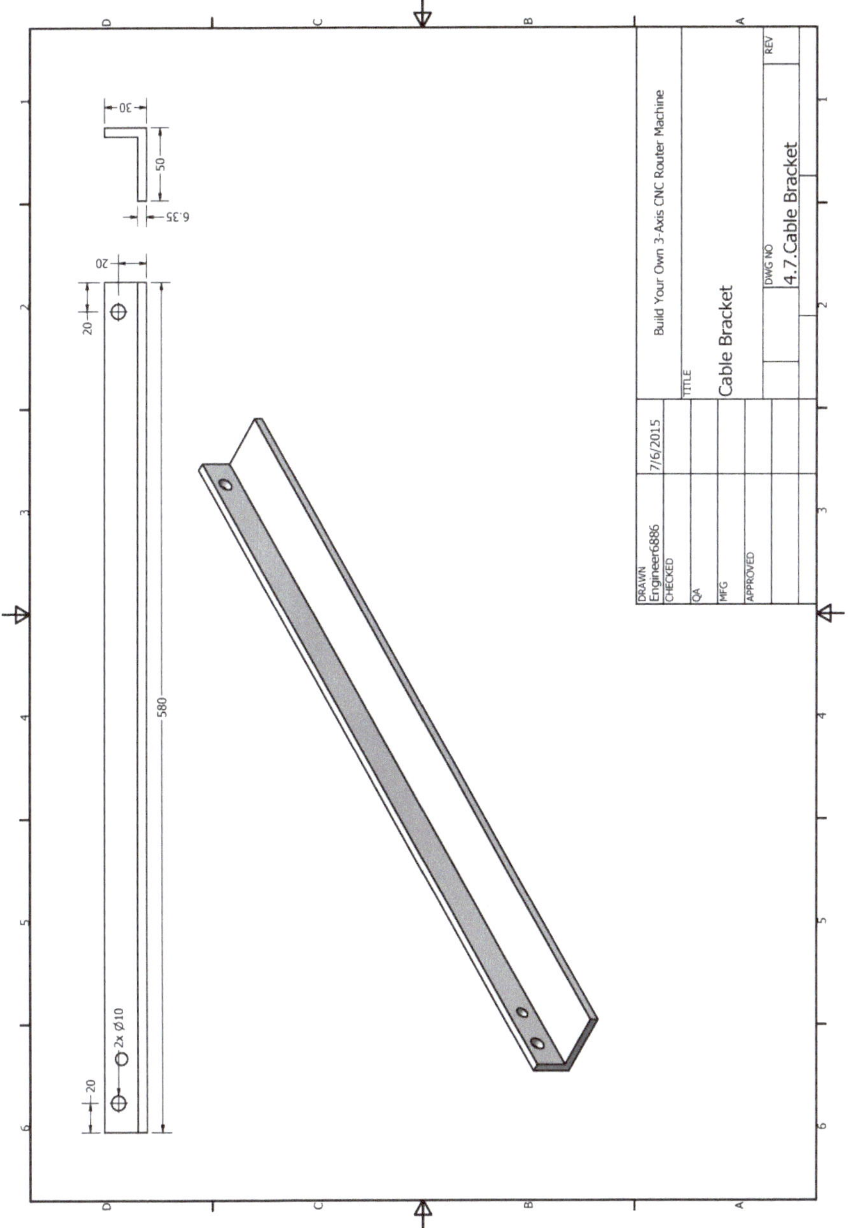
30
50
6.35
20
20
580
2x Ø10
20
Build Your Own 3-Axis CNC Router Machine
TITLE
Cable Bracket
DWG NO
4.7.Cable Bracket
REV
DRAWN
Engineer6886
7/6/2015
CHECKED
QA
MFG
APPROVED

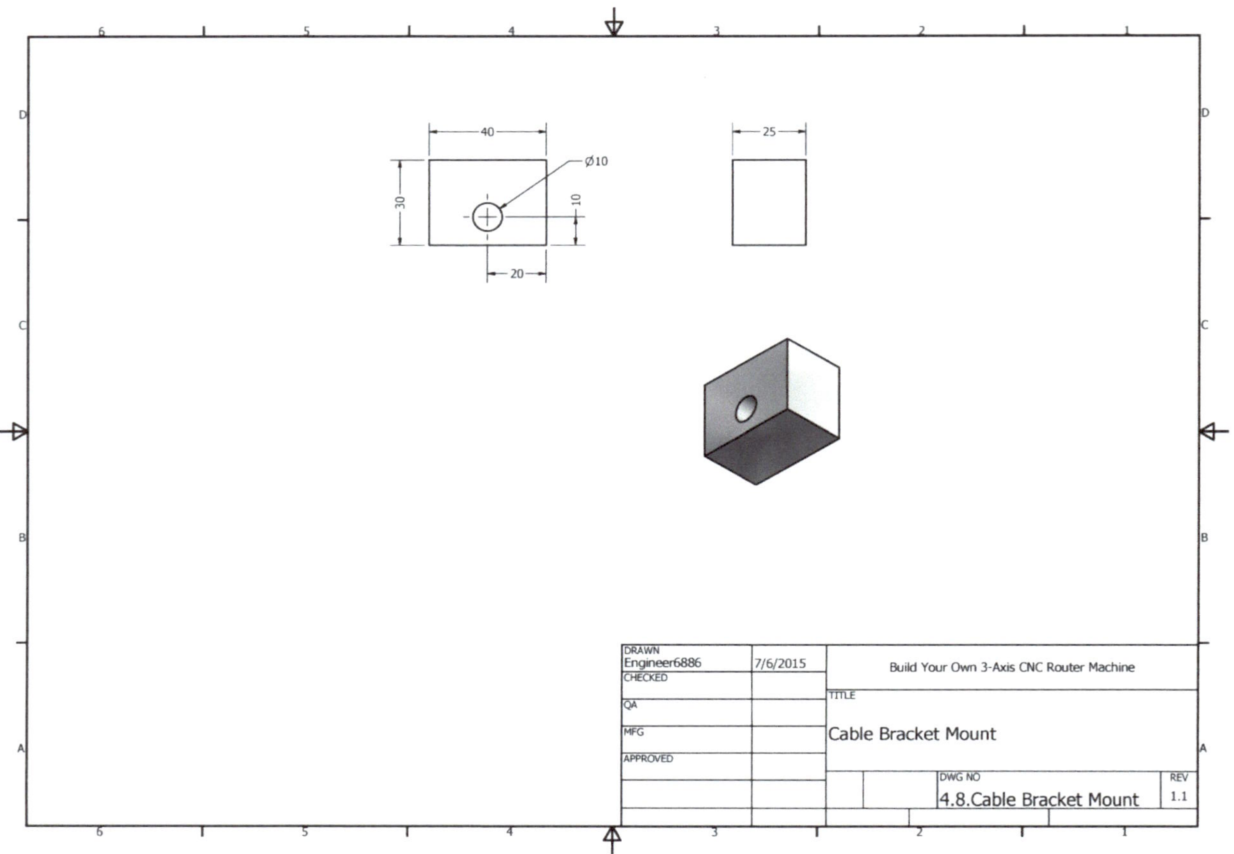
40
Ø10
25
30
10
20
DRAWN
Engineer6886
7/6/2015
Build Your Own 3-Axis CNC Router Machine
CHECKED
TITLE
QA
MFG
Cable Bracket Mount
APPROVED
DWG NO
4.8.Cable Bracket Mount
REV
1.1
1
2
3
4
5
6
A
B
C
D

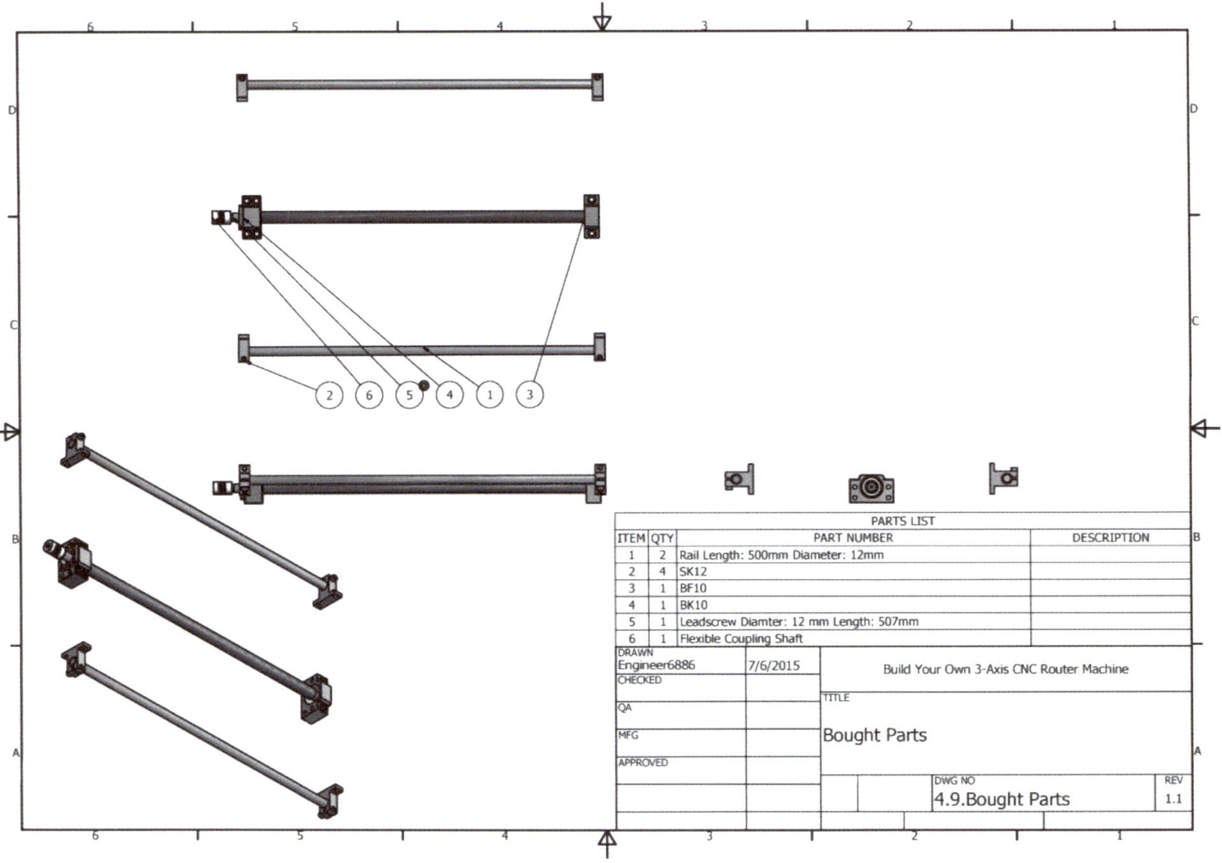

PARTS LIST
ITEM | QTY | PART NUMBER | DESCRIPTION
1 | 2 | Rail Length: 500mm Diameter: 12mm
2 | 4 | SK12
3 | 1 | BF10
4 | 1 | BK10
5 | 1 | Leadscrew Diamter: 12 mm Length: 507mm
6 | 1 | Flexible Coupling Shaft
DRAWN
Engineer6886
7/6/2015
CHECKED
QA
MFG
APPROVED
Build Your Own 3-Axis CNC Router Machine
TITLE
Bought Parts
DWG NO
4.9.Bought Parts
REV
1.1

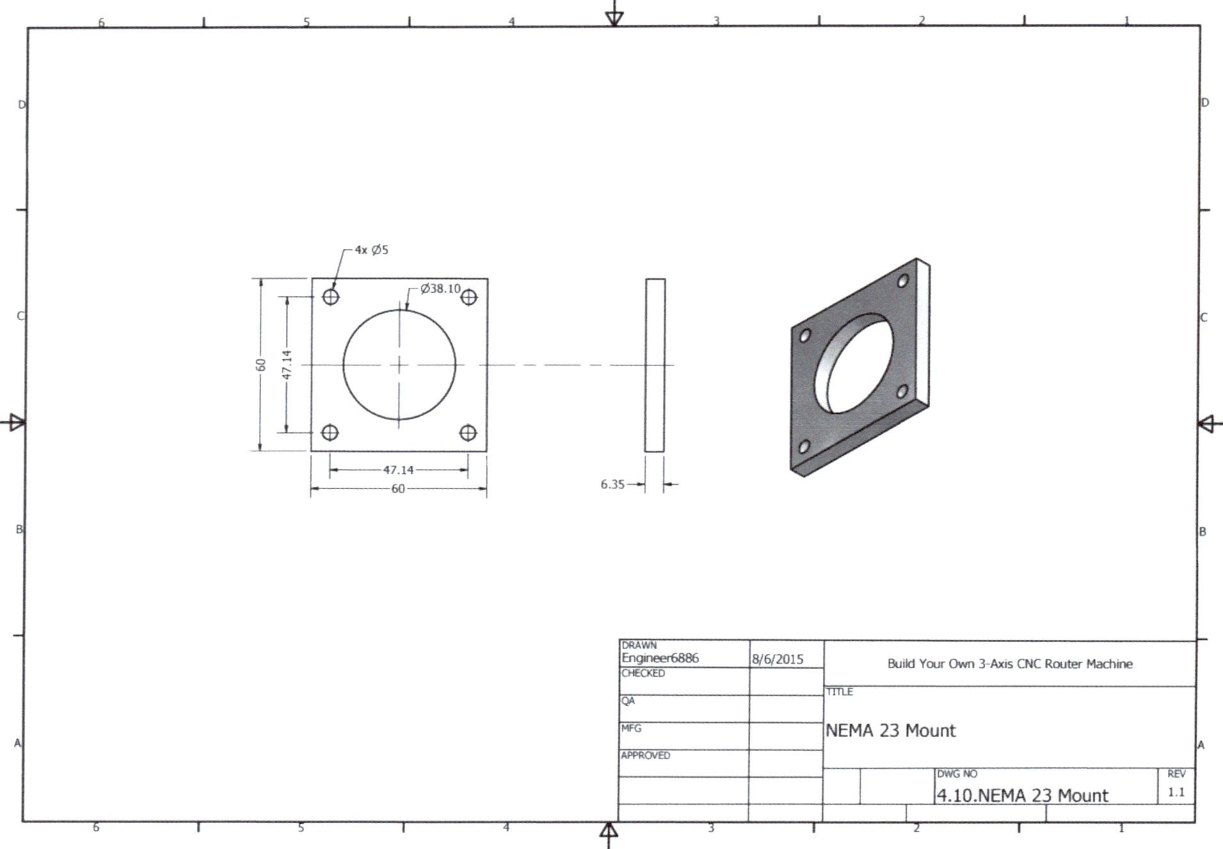
4x Ø5
Ø38.10
60
47.14
47.14
60
6.35
DRAWN
Engineer6886
8/6/2015
CHECKED
QA
MFG
APPROVED
Build Your Own 3-Axis CNC Router Machine
TITLE
NEMA 23 Mount
DWG NO
4.10.NEMA 23 Mount
REV
1.1

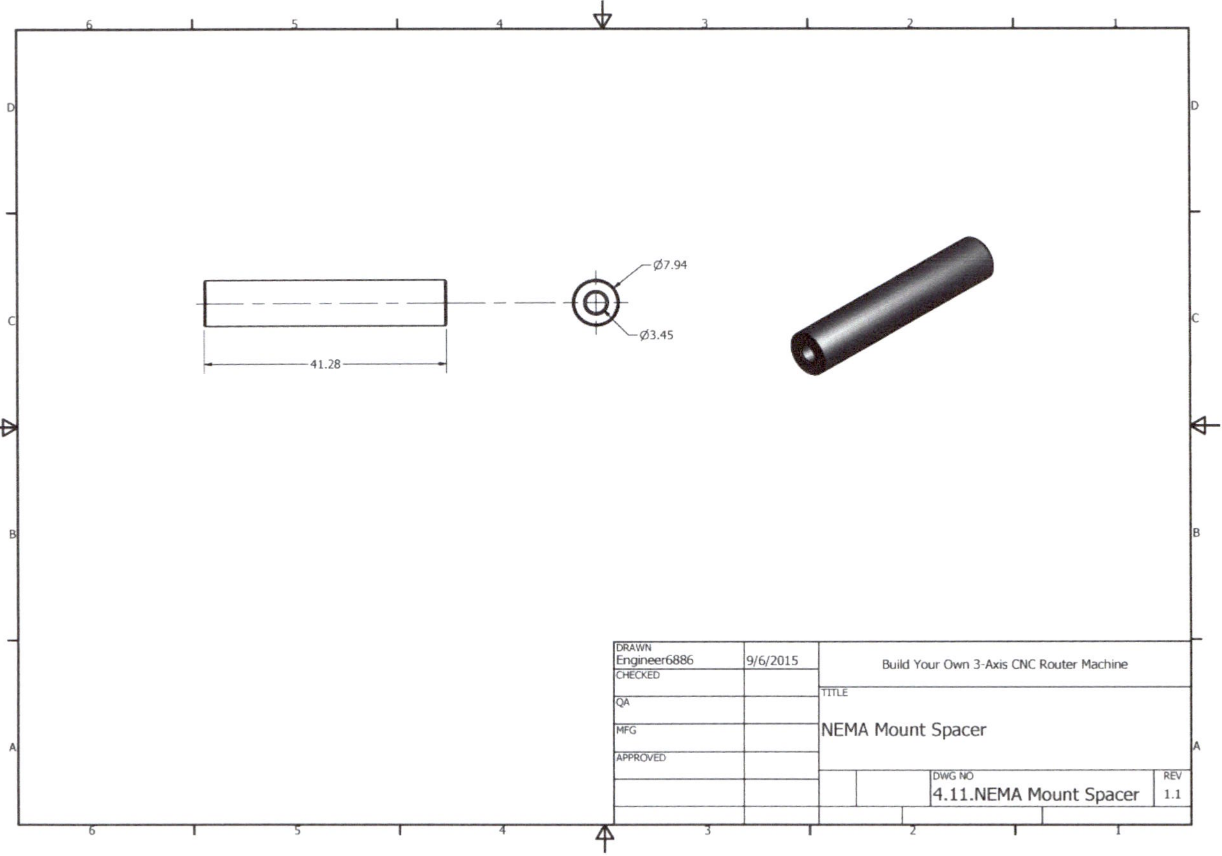
41.28
Ø7.94
Ø3.45
DRAWN
Engineer6886
9/6/2015
CHECKED
QA
MFG
APPROVED
Build Your Own 3-Axis CNC Router Machine
TITLE
NEMA Mount Spacer
DWG NO
4.11.NEMA Mount Spacer
REV
1.1

Fabricating the metal parts

The most efficient method of fabricating the metal parts is to do each type of operation all at once. More in-depth description of each step are on the following pages. The following fabrication sequence was used here:

1. Cut all the pieces to their overall length. If necessary, file the cut ends of the parts while holding them in vise, to clean them up and make them as perpendicular as possible to the part.
2. Lay out all locations of holes and any secondary cuts with a marker and scribe.
3. Center punch all holes. Use a spring loaded centerpunch for accuracy and speed.
4. Drill all holes. It may be quicker to drill all same size holes at once.
5. Drill counterbore holes. It may be easier to drill the primary hole, keep the part clamped, change the bit and then drill the counterbore. This will maintain concentricity.
6. Tap all threaded holes.

Construction Schedule

The construction of the machine starts with the build of the Carriage Table. It is the easiest part of the whole assembly and the builder can have an idea of what it is coming next.

Second assembly is the Z – Axis. The carriage table slides in Z–Axis providing the linear perpendicular movement of the spindle.

The construction continues with the Y–Axis and X–Axis construction. Those 2 Axes provide the plane motion to the spindle.

Last, the construction is completed with the construction of the electronics box.

Construction – Carriage Table

For the complete manufacturing of the carriage table, the builder will need to manufacture the parts 1.1 and 1.2 and also buy the following parts (1.3):

- ✓ 16x M4x14 Socket Head Bolts
- ✓ 2x M5x20 Socket Head Bolts
- ✓ 6x M4x25 Socket Head Bolts
- ✓ 6xM4x0.7 Prevailing Torque Nuts
- ✓ 4x SC12 Linear Bearings
- ✓ 1x SF12 Ball Nut

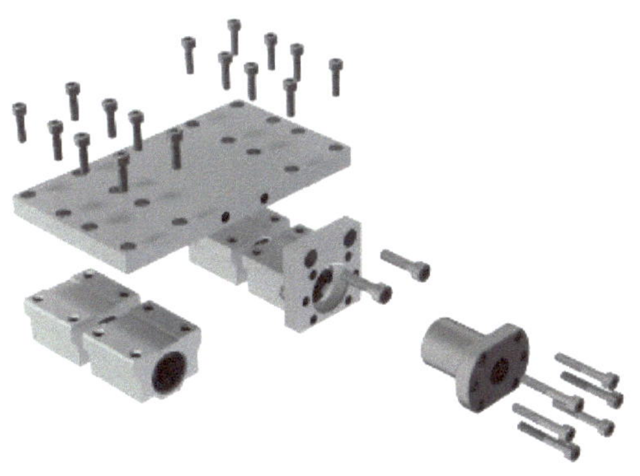

First, the part 1.1 has to be manufactured. After the cut to the final dimension, the holes are marked with a spring loaded centerpunch.

Then the holes are drilled and counterbored through the entire thickness of the material. The two side holes for the Ball Nut Mount must be tapped to the desired thread. The part should look like the photo below.

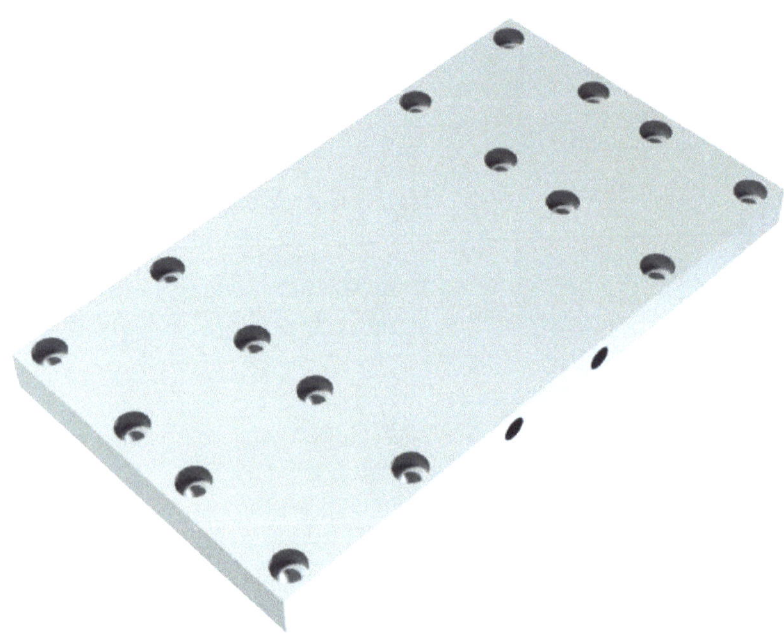

Next is the part 1.2 to be manufactured. After its cut to the final dimensions, the holes are marked again with a spring loaded centerpunch and drilled.

Builder can decide to tap the holes for the Ball Nut Mount or to not tap and use Prevailing Torque Nuts to secure the SFU 12 to its Mount. The finished part can be seen in the photo below.

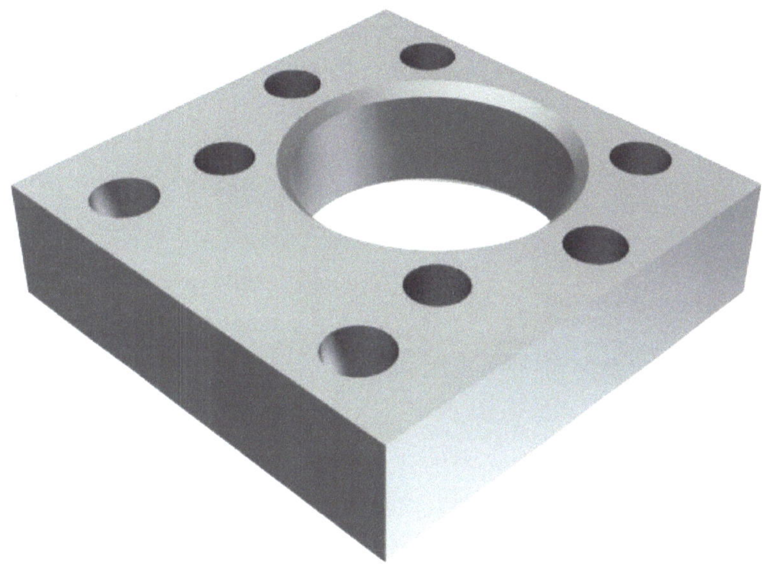

Procedure of assembly (Carriage Table)

1. Bolt the four (4) SC12 Linear Bearing to the Carriage Table with the M4x14 Socket Head bolts.
2. Bolt the Ball-Nut Mount to the Carriage Table with the M5x20 Socket Head bolts.
3. Bolt the SF12 Ball-Nut to the Ball-Nut Mount with the M4x25 Socket Head bolts and M4x0.7 Prevailing Torque Hexagon Nuts

The Carriage Table assembly is completed and should look like the photo below.

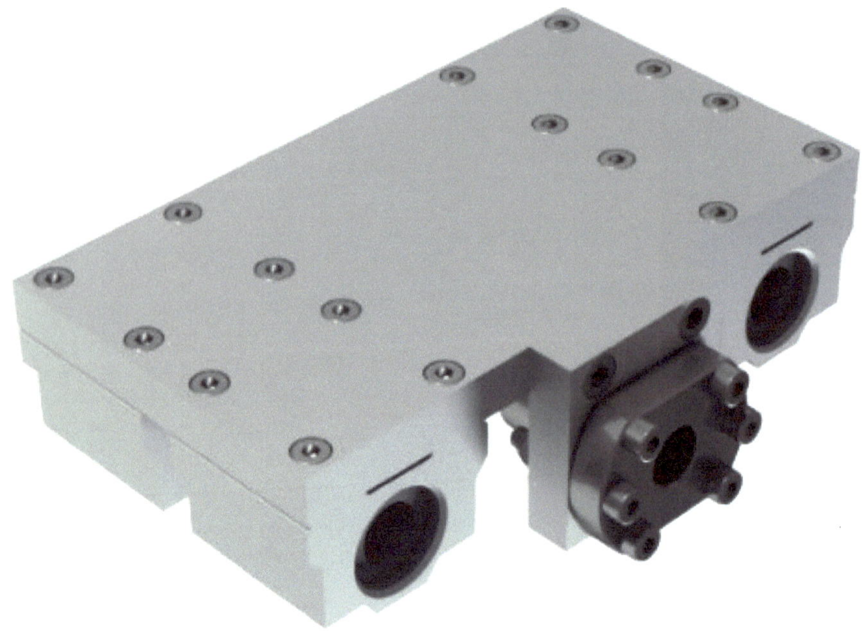

Next is the construction of the Z Axis. The Carriage Table that has been already made slides into the Z Axis. The Z Axis consists again with parts that needs to be manufactured and parts that need to be bought.

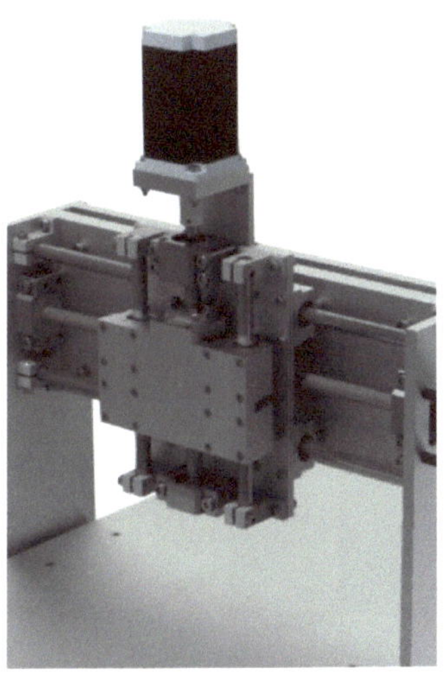

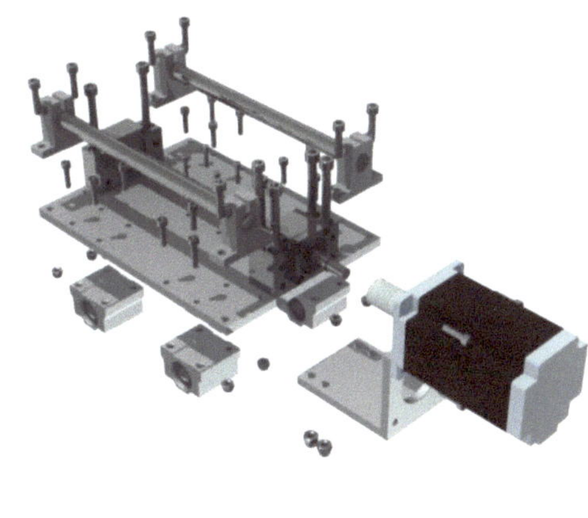

For the complete manufacturing of the Z - Axis, the builder will need to manufacture the parts 2.1, 2.2 and 2.5 and also buy the following parts (2.3):

- ✓ 16x M4x14 Socket Head Bolts
- ✓ 2x M5x20 Socket Head Bolts
- ✓ 6x M4x25 Socket Head Bolts
- ✓ 6x M4x0.7 Prevailing Torque Hexagon Nuts
- ✓ 4x SC12 Linear Bearings
- ✓ 1x SF12 Ball Nut
- ✓ 4x Sk12 Rail Mounts with Rails
- ✓ 1x BK & BF Mount with Leadscrew and Coupling

First, the part 2.1 has to be manufactured. After the cut to the final outer dimensions, the holes are marked with a spring loaded centerpunch. Then the holes are drilled through the thickness of the material. Counterbore holes are not needed on that part.

The finished part should look like the photo below.

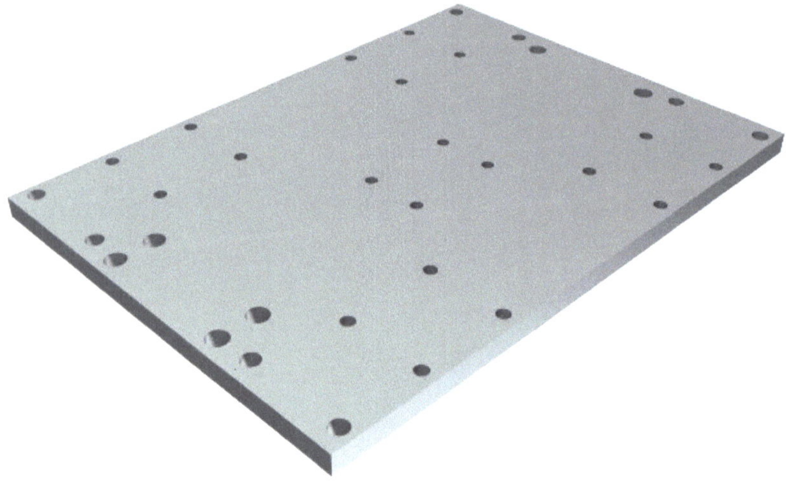

Next is the part 2.2 to be manufactured. For that part, a flat aluminum plate can be used, drilled and then bent or use an angled piece of aluminum and drill each side separately.

The finished part should look like the photo below.

Last part to be constructed is the part 2.5. That part can be constructed by cutting in half a rectangular aluminum section. The finished part can be seen below.

Procedure of assembly (Z-Axis)

1. Bolt the four (4) SC12 Linear Bearing to the Z-Axis Back Plate with the M4x14 Socket Head bolts.
2. Bolt the SF12 Ball-Nut to the Ball-Nut Mount with the M4x25 Socket Head bolts and M4x0.7 Prevailing Torque Nuts
3. Bolt the Ball-Nut Mount to the Z-Axis Back Plate with the M5x20 Socket Head bolts.
4. Bolt the four (4) SK12 Rail Mounts to the Z-Axis Back Plate with the M5x20 Socket Head bolts.
5. Insert one SFU 12 Ball-Nut to the Leadscrew and thread it to the middle of the shaft.
6. Place the Leadscrew into the BF10 and BK10 Mounts.
7. Bolt the BF10 and BK10 Mounts to the Z-Axis Back Plate.
8. Slide 4 SC12 (extra four) into the Rails and insert the rails to the SK12 Rails Mounts.
9. Bolt the Stepper Motor Plate to the Z-Axis Back Plate.

10. Bolt the Stepper Motor to the Stepper Motor Plate together with the Flexible Coupling Shaft.

The Z-Axis assembly is complete and should look like the photo below (In the photo the extra four (4) SC12 are not shown).

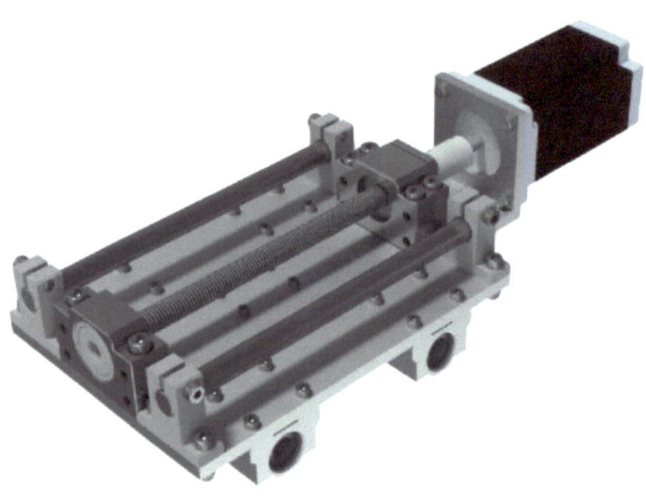

Next is the construction of the Y-Axis. The Carriage Table and the Z-Axis that have been already made, slide on the Y-Axis providing the "lateral" movement to the spindle.

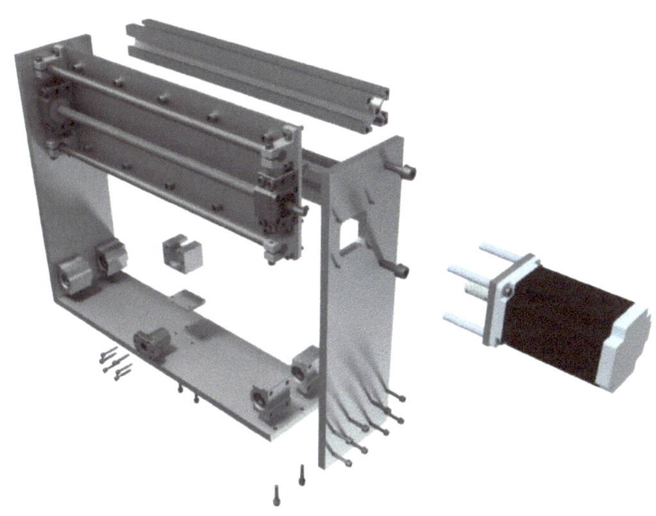

For the complete manufacturing of the Y-Axis, the builder will need to manufacture the parts 3.1, 3.2, 3.3, 3.4, 3.5, 3.8, 3.9, 3.10 and 3.11 and also buy the following parts (3.6 and 3.7):

- ✓ 10x M6x16 Socket Head Bolts
- ✓ 12x M5x20 Socket Head Bolts
- ✓ 4x M5x20 Socket Head Bolts
- ✓ 6x M6x50 Socket Head Bolts
- ✓ 2x M6x20 Socket Head Bolts
- ✓ 4x M8x45 Socket Head Bolts
- ✓ 30x M4x20 Socket Head Bolts
- ✓ 4x SC12 Linear Bearings
- ✓ 1x SF12 Ball Nut
- ✓ 1x complete set of Rails, Sk Mounts, BF, BK and Leadscrew
- ✓ T-Nuts for the aluminum Profile

All the parts for the Y-Axis will be constructed with the same sequence. After the cut to the final dimensions, the holes are marked with a spring loaded centerpunch on one side and then drilled through the entire thickness of the material. Counterbore holes are not needed for that parts.

Keep in mind that in the part 3.3 must be made an opening cut for the BK Mount placement. That can be done with many ways but the cheaper method is with stich drilling and filing (drill many holes on the outline of the opening cut and then file it to the final dimensions).

The finished parts should look like the photos below.

Part 3.1.Y-Axis Back Plate

Part 3.2.Aluminum Profile 40x40

Part 3.3.Right Side Plate

Part 3.4.Left Side Plate

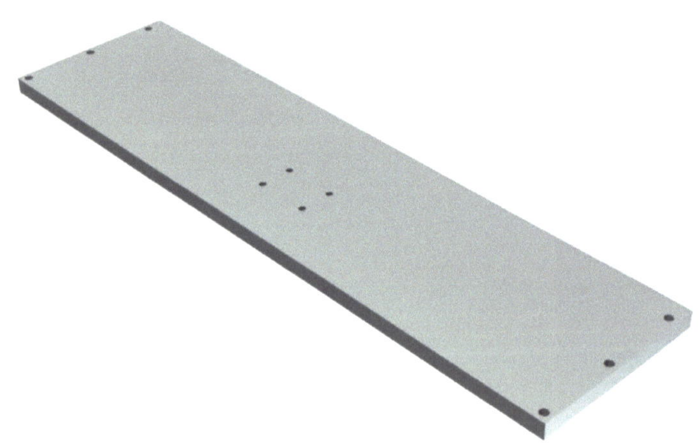

Part 3.5.Lower Plate

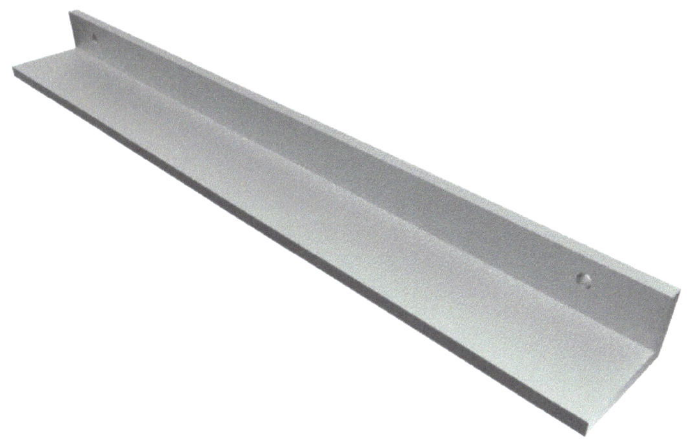

Part 3.8.Cable Bracket

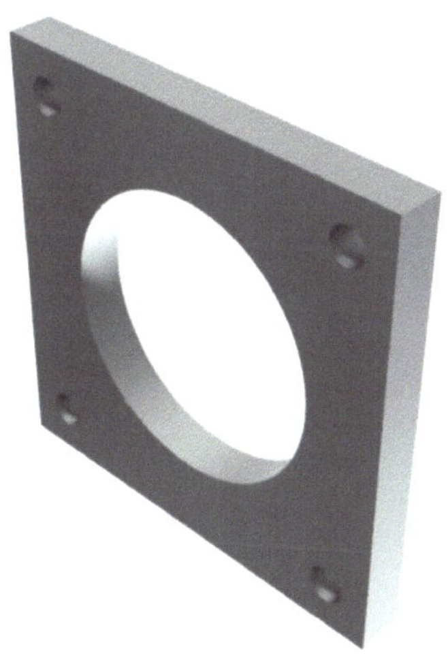

Part 3.9.NEMA 23 Mount

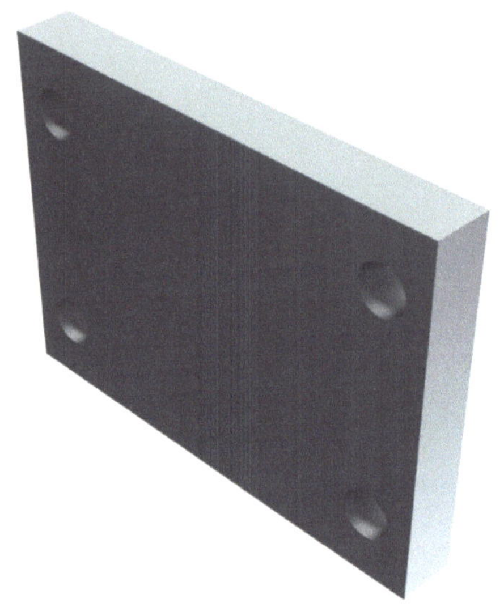

Part 3.10.Ball-Nut Mount Spacer

Part 3.11.NEMA 23 Mount Spacer

1. Bolt the two (2) parts 3.2.Aluminium Profile 40x40 to the part 3.1.Y-Axis Back Plate with the M6x16 Socket Head bolts.
2. Bolt the SK12 Rail Mounts to Y-Axis Back Plate with the M5x20 Socket Head bolts
3. Bolt the BF and BK Mount to the Y-Axis Back Plate with the M6x50 Socket Head bolts.
4. Bolt the parts 3.3 and 3.4 (side plates) to the assembly above with M8x45 Socket Head bolts.

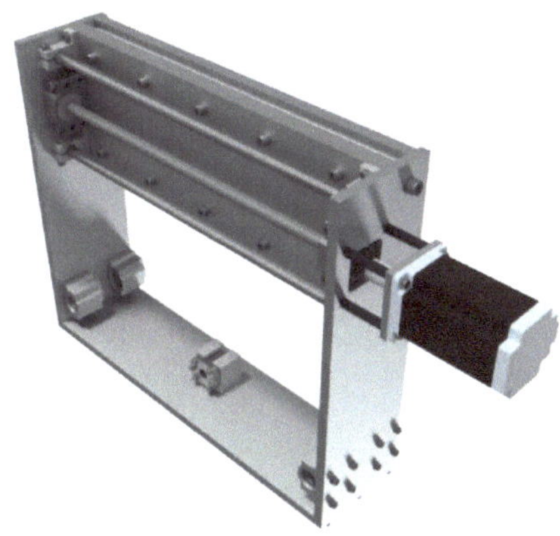

5. Bolt the part 3.5.Lower Plate to the assembly with M5x25 Socket Head bolts.
6. Bolt the four (4) SC12 Linear Bearings to the two (2) side plates with M4x20 Socket Head Bolts.
7. Bolt the Ball-Nut Mount together with Ball-Nut to the Lower Plate with M4x20 Socket Head Bolts.
8. Bolt the part 3.9.NEMA 23 Mount to the Right side Plate with M4x20 Socket Head Bolts.
9. Bolt the part 3.8.Cable Bracket to the rear of the whole assembly.
10. Bolt the Stepper Motor to the NEMA 23 Mount together with the Flexible Coupling Shaft.

Next is the construction of the X Axis. All the sub-assemblies made until now, slide on the X-Axis to provide the 3-Axis movement of the spindle.

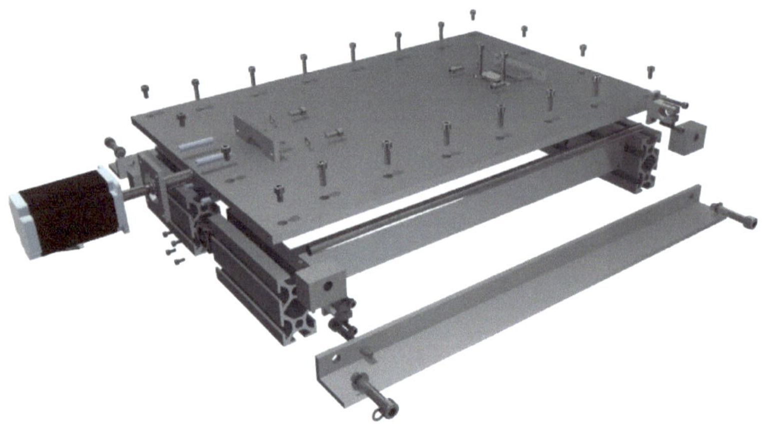

For the complete manufacturing of the X - Axis, the builder will need to manufacture the parts 4.1, 4.2, 4.3, 4.4, 4.5, 4.6, 4.7, 4.8, 4.10 and 4.11 and also buy the following parts (4.9):

- ✓ 8x M6x10 Socket Head Bolts
- ✓ 16x M5x20 Socket Head Bolts
- ✓ 6x M6x40 Socket Head Bolts
- ✓ 8x M6x16 Socket Head Bolts
- ✓ 8x M5x30 Socket Head Bolts
- ✓ 4x M10x50 Socket Head Bolts
- ✓ 1x M8x60 Socket Head Bolts
- ✓ 1x complete set of Rails, Sk Mounts, BF, BK and Leadscrew
- ✓ T-Nuts for the aluminum Profile

First, the part 4.1 has to be manufactured. After the cut to the final dimensions, the holes are marked with a spring loaded centerpunch. Then the holes are drilled through the thickness of the material and later counterbored.

The finished part should look like the photo below.

All the other parts for the X-Axis will be constructed with the same sequence. After the cut to the final dimensions, the holes are marked with a spring loaded centerpunch on one side and then drilled through the thickness of the material. Counterbore holes are not needed for that parts.

The finished parts should look like the following photos.

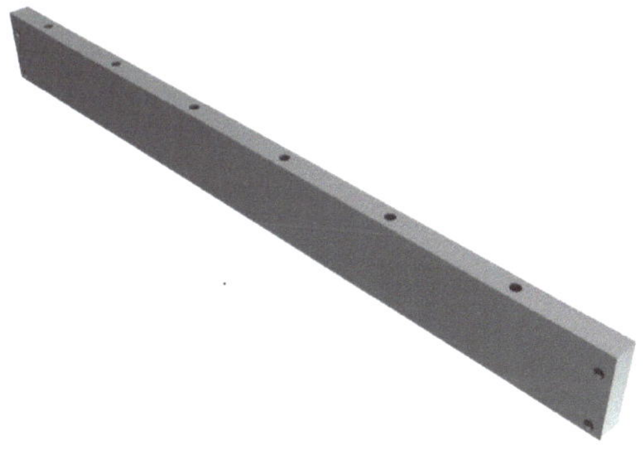

Part 4.2.Table Side Plate

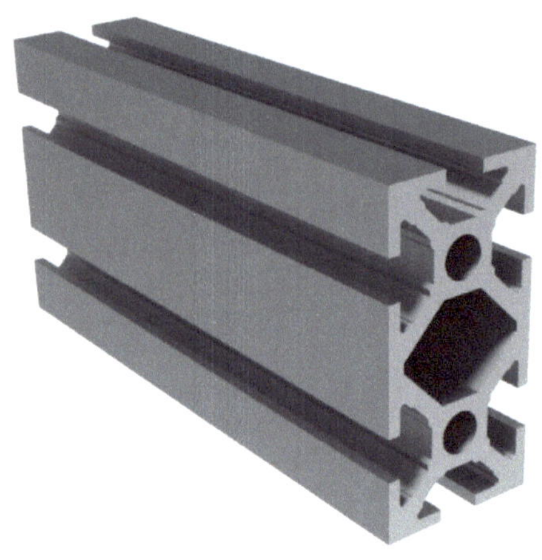

Part 4.3.Aluminum Profile 80x40

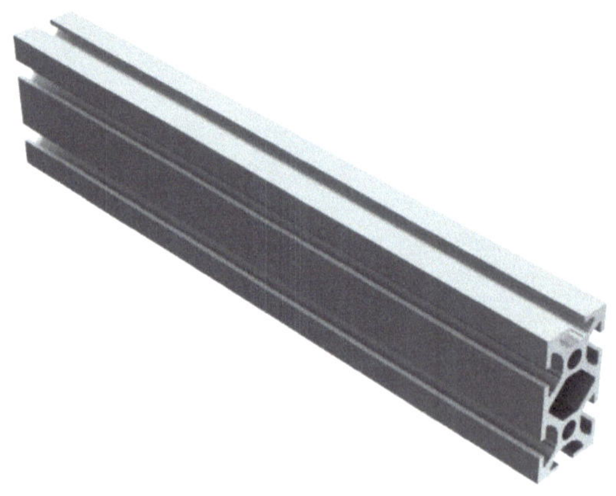

Part 4.4.Aluminum Profile 80x40

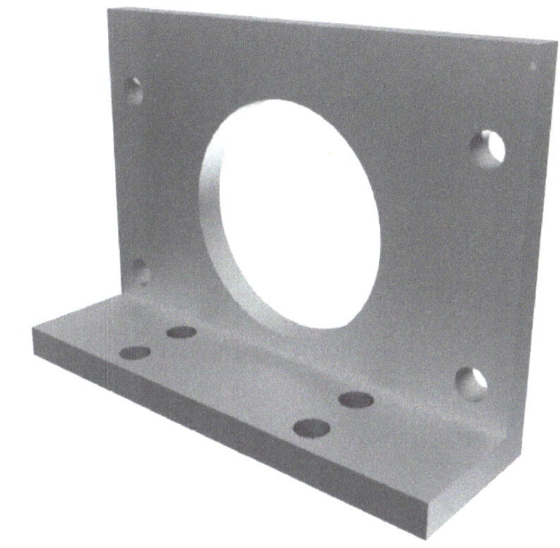

Part 4.5.BK Mount

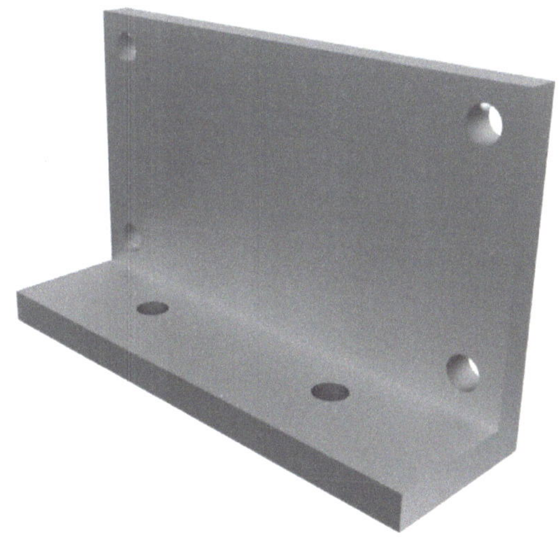

Part 4.6.BF Mount

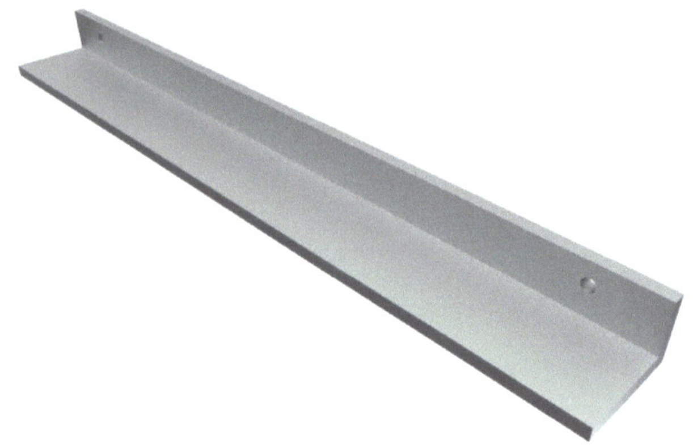

Part 4.7.Cable Bracket

Part 4.8.Cable Bracket Mount

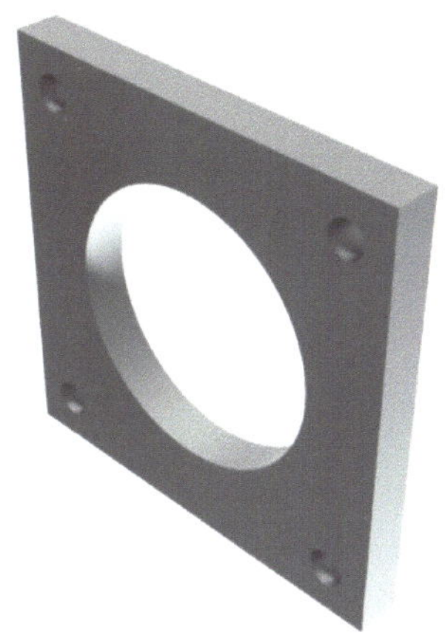

Part 4.10.Nema 23 Mount

Part 4.11.NEMA 23 Mount Spacer

1. Insert the leadscrew to its BF and BK mounts.
2. Bolt the BF and BK mount to their angle mounts respectively (parts 4.5 and 4.6).
3. Bolt the aluminum profiles 80x40 (parts 4.3 and 4.4) to the BF and BK Mounts respectively.
4. Bolt the 4 Cable Bracket Mounts (part 4.8) to each "free" side of the aluminum profiles.
5. Bolt the two sides plates (part 4.2)
6. Place the two Cable Brackets, one on each side of the Router
7. Insert the two rails to their SK12 mounts and bolt the mounts to the side plates.
8. Bolt the NEMA23 Stepper mount with its mount to the BK mount.
9. Bolt the Base Plate (part 4.1) in place.

Cable Chains Mounting

Cable chains are used to "hide" and also protect the cables that connect the stepper motors and spindle with the drives and breakout board. The best way to not destroy your cables when the machine is working is to use those cable chains.

The photo below shows a mounting method of the cable chains. You can use this method or change the orientation of the chains.

The basic principle is that you need two chains, one for the X-Axis and one for the Y-Axis. All the cables can be routed inside of those two chains.

The mounting can be done easily with just a bolt on each side of the chains.

Electronics Box Construction

The electronic box can be kept as simple as possible or as complex the builder wants to be. Also, it can be kept as small as possible or very large for better cooling. The construction of an electronics box is not in the scope of the drawings. It's up to each builder to mount the electronics wherever he/she wants. The only thing for sure is that an electronics box is **<u>needed</u>** 100%.

Switches like e-stops, on/off, probe tools can be mounted on the external sides of the box making the use of the machine easier.

In the photo below, the box kept as small and simple as possible.

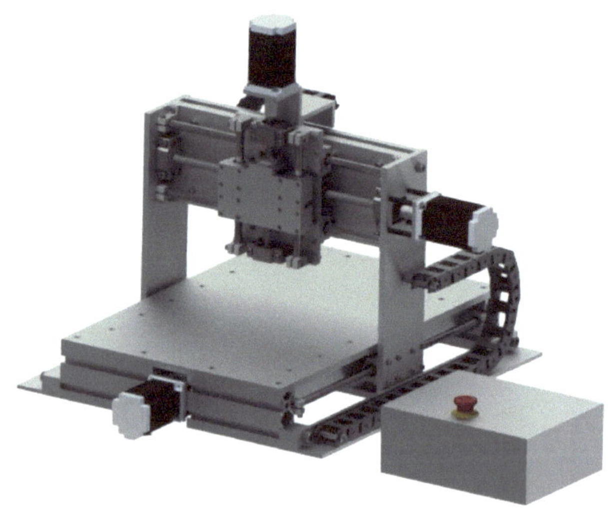

Limit Switches Placement

The positions showed below are just some of many possible placements that the builder can choose. Choose the position of the switches for your machine in a way to maximize the axis travel, to have no possibility for the wiring to be tangled with the carriage table and the switches can be activated from the linear bearings.

Be sure to connect the wiring to the correct lugs, so that the limit switches will be operating in the normally closed configuration. The image below shows the upper and lower Z-Axis limit switches. Place them as wide as possible in order to have more travel for the carriage table. A spacer should be used below each limit switch for the mounting, so it will have enough clearance from the bolts on the back plate.

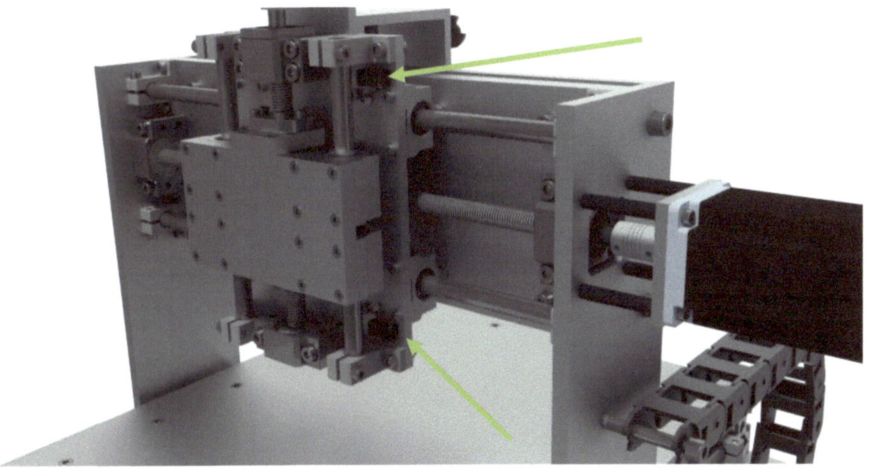

The X-Axis limit switches are placed with the same logic. Holes can be opened directly behind the switches for the wiring and the connection with the Breakout Board. A photo of the X-Axis limit switches is shown below.

Last, the limit switches of the Y-Axis are shown below. Limit Switches must be wired in *normally closed* configuration.

Each change – breakage on the wiring must stop the machine from running.

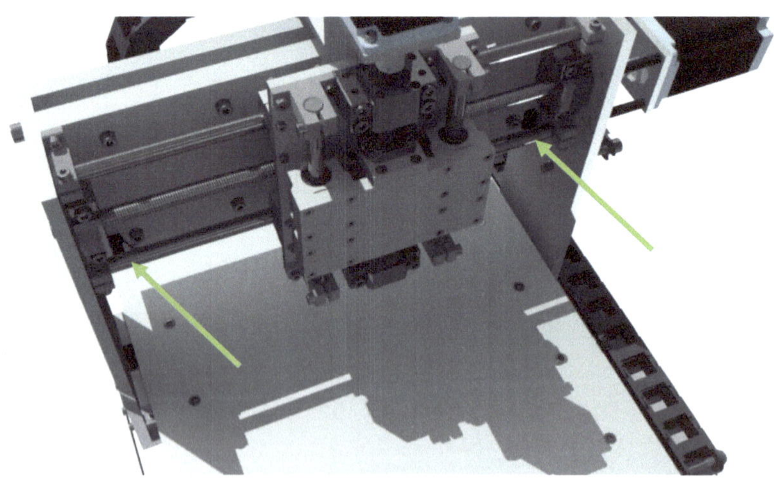

Electronics

Due to the complexity of choosing individual components that will function together, it is highly recommended to purchase a pre-packaged kit from a supplier that includes all of the electronics as a matched collection.

The basic principles of what is included in an electronic set, are:

- ✓ A computer to send motion data to a hardware component called "motor drive"
- ✓ A power supply to provide the required voltage and current to the motors
- ✓ Electronic "motor drives" that forwards the motion data to the stepper motors at the required voltage/current
- ✓ The stepper motors

This is the general flow of information from the computer to the motors. In addition, there are typically hardware components to provide data feedback from the machine to the computer. All systems should be equipped with limit switches at the end of each axis travel, to provide safety to both the machine and operator. More sophisticated machines may have feedback sensors that give more accurate control of the motor positioning.

Motors

Most home-built machines are controlled by stepper motors. These are simple type of DC motors that require a pulse of electricity to move the motor one step. A "typical" stepper motor has 200 steps per revolution (without micro-stepping), so to cause continuous rotation in a stepper motor it requires a fast steam of discrete electrical pulses. The frequency of the pulses will determine the motor speed.

These motors are easy to electronically control via computer, and relative inexpensive, but they do have some drawbacks. One is that there is a possibility of "losing steps", under a load. This happens due to the stream of electrical steps pulses continuing to flow to the motor even through it is temporarily being prevented from moving. Since the number of steps required to move the machine is very high, a very small number of missed steps may not have any noticeable impact on the finished part. On the other hand, enough missed steps may be catastrophic. In the best case, it may result in a less than perfect and at worst it may result in machine collision, since after the event that causes lost steps the machine location is not corresponding to where the software thinks it should be. Missed steps is a problem with stepper motors because they typically lack any sort of feedback mechanism. They simply do as they told to do and the control software has no way of recognizing any error that may occur or a way to correct the motion.

Drives

All motors that are used for motion control require some type of electronic drive board to control them. Drive boards take a variety of arrangements. They may incorporate control for several axes on a single board or may be configured as an individual board of each axis. The advantage of a separate board per axis is that they can be replaced individually in case of damage.

All drive boards do essentially the same thing. They receive input signals from the control software, which are low in voltage and current and output these signals to the motors with higher voltages and currents that they require for operation. As such, they mediate between the computer and machine. Their in-between position also allows them to handle signal inputs for additional functions such as emergency stop buttons and limit switches. Most drive boards are vulnerable to any errors in miss-wiring. Incorrect connections or braking a connection to the motor while under power, can cause an immediate destruction of the electronics on the board.

Micro-stepping

Another function of many drives is that they break up a number of steps per revolution that are required at the motor into a greater number. So for instance a drive may have "1/8" or "8x" micro-stepping, which would increase the number of steps per revolution from its original 200 to 1600. This is an advantage because it increases the resolution of the system and provides finer control over the movement of the machine.

Power Supply

A transformer type power supply device is matched to the needs of the drive boards and motors. It is worth noting that many boards that operate by "pulse width modulation" perform most efficiently at the upper limits of voltage that they can handle. In other words, running them in low voltage will not provide any additional protection for the drive board. Stepper motors also commonly require many times more voltage than their ratings may indicate. For instance, a stepper motor that is designated as a 2.5 Volt motor may require a 24 Volt power supply to efficiently power it.

Breakout Board

Simple drive boards such as those that have multiple axes self-contained on a single board may be designed for direct connection to the computer's parallel port via a standard cable. When using multiple drives that each control an individual axis, an additional piece of hardware called a "breakout board" may be required. This is a device that connects to the computer via a cable (parallel port) and provides multiple connections to allow wiring to the drives, emergency stops switches, limit switches, spindle control relays etc. These boards also provide an added layer of protection between the higher voltage drive boards and the vulnerable low voltage computer. They do this through optically isolated connections.

Additional Switches

An emergency stop button should be a part of every system. It is typically a large red button with a mushroom-shaped head that provides an immediate way of shutting down the machine I case of emergency. It should be placed in a location that is easily accessible while operating the machine. It can be wired to shut down all axes of machine movement and can also be wired to shut down power to the spindle rotation. If at all possible your e-stop should be wired in this manner.

The other switches that should be wired to the system are limit switches. These are placed at the end of each axis's movement, so a 3-Axis machine will have 6 limit switches. These switches will stop the motion of the machine if unexpectedly reaches the end of the travel axis. This can prevent serious damage to the machine as well as guarding against personal harm from broken cutters. These switches can be used also as homing switches. These are used to return the machine automatically to its home XYZ position. Most control software can be configured to use the switches in this manner.

Electronics Setup

Because of the complexity of the electronics used in CNC machines, it is better to use a ready kit from a manufacturer that contains every electronic part needed to make the machine run.

In the next pages, the default electronic parts that used in the prototype machine are listed. There will be a description for every part together with the correct connection that has to be made.

For the stepper motors wiring, 18AWG with 4 cores shielded cable **SHOULD** be used. In general it is better to use shielded cables for all the wiring in the machine.

In the next page, the complete cable harness can be seen. Keep in mind that all of those electronic parts, except stepper motors and limit switches, must be kept in the "electronics box" next to the CNC Router.

The emergency button (e-stop) should be placed in a place that the user should be able to reach easy in order to stop the machine if something's going other way than it should.

In the next page, the drawing of the complete machine harness can be seen.

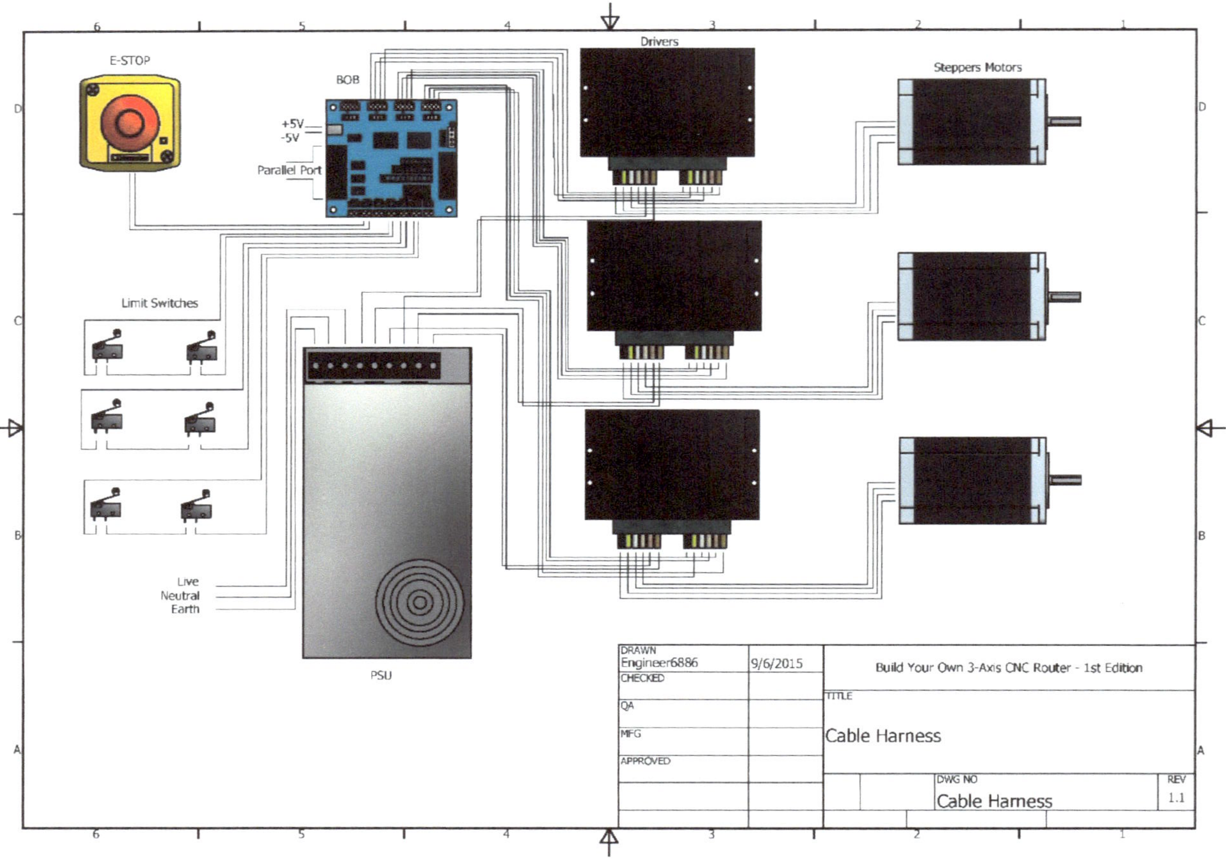
E-STOP
BOB
+5V
-5V
Parallel Port
Drivers
Steppers Motors
Limit Switches
Live
Neutral
Earth
PSU
DRAWN
Engineer6886
9/6/2015
CHECKED
QA
MFG
APPROVED
Build Your Own 3-Axis CNC Router - 1st Edition
TITLE
Cable Harness
DWG NO
Cable Harness
REV
1.1

Most of the Power Supply Units when bought are not coming with a pre-installed plug to 220AC. It is very easy to make that connection. An already extension cord can be cut and used to power up the PSU. The only "tricky" thing is to identify the 3 cores inside the cable.

Those cores are:

- ✓ Live
- ✓ Neutral
- ✓ Earth

As soon as those cores are identified, a stable connection has to be made between them and the "screw-plugs" on the PSU.

Before you connect anything else to the PSU, connect the PSU to the 220AC power outlet and with a voltmeter check the **output voltage** from the PSU.

The output voltage should be in specific voltage in order to connect the drives to PSU. The manufacturer of the drives should mention the input voltage for the drives. A wrong connection in that phase (more voltage to drives) can destroy the drives. Usually that price is 36V. If for any reason that's not the price that the voltmeter is showing, you can correct it by using the potentiometer on the left side of the PSU (screw it or unscrew it) until the output voltage is 36V.

The cooling fan of the PSU should not work all the time. It will work only if the temperature raise to specific limit.

The PSU that used on the prototype router is a 350W power supply and 36V output.

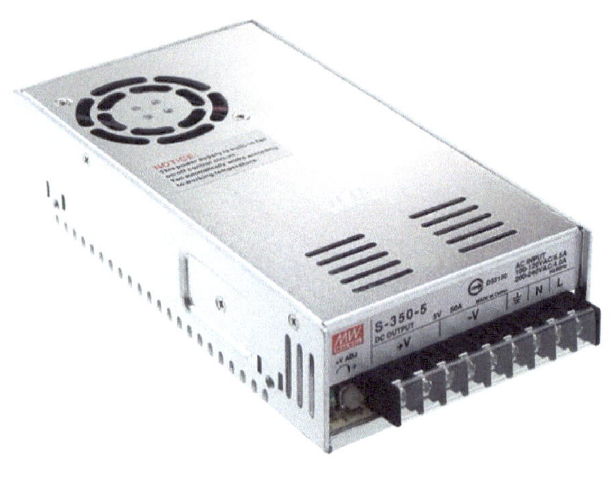

Typical PSU used in CNC machines

As soon as the output voltage is set correct, you can continue the wiring from the drives to the PSU. The drive that was used in the prototype CNC Router Machine is the **DM542A** and for that drive the wiring will be explained in the following pages.

In order to make the drives to work you must wire them to the PSU from the "High Voltage" side and more specific connect the "DC- and DC+" to the "-V and +V" on the PSU. That job has to be done for all drives.

The wiring between the drives and BoB can be done through the signal ports on the upper side of the DM542A. Below you can see the breakout board used in the prototype CNC Router. The code name of that BoB is **HY-JK02-M**.

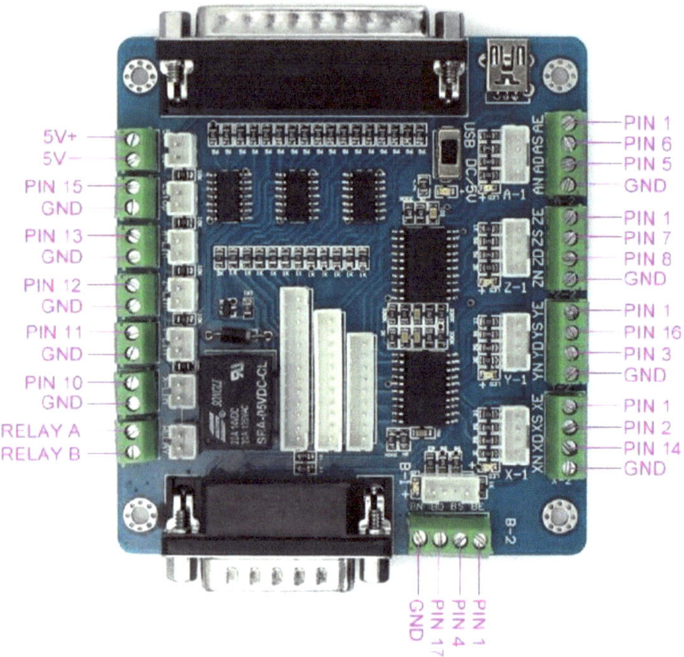

That BoB can handle up to 5 CNC axes. As seen in the photo there are a lot of pins and ports that we are going to explain in the next sentences below.

Starting from the right side of the BoB, can be seen 4 “packets” of pins. Each pack of pins is for an individual Axis (X, Y, Z, A) and the lower right pack is for the B Axis.

Each pack consist of 4 ports. The name of the pins are listed below:

- ✓ N for Ground
- ✓ S for Pulse
- ✓ D for Direction
- ✓ E for Enabled

In order to make a correct wiring, all the "-"ports from the "signal" section on all DM542A must connect to the "N=Ground" port on each section of pack on the breakout board.

Then the ENBL+ post from DM542A connects to the "E" port on Bob, the DIR+ port from DM542A connects to the "D" port on breakout board and the PUL+ port from DM542A connects to the "S" port on BoB.

That job has to be done for each drive in order to have 3 workable drives and later stepper motors.

Next connection is between <u>Bob to its Power Outlet</u> and to <u>Personal Computer</u>.

The BoB can power up with 3 different methods. Each method is 100% correct and it is just which solution is more practical for the builder. BoB needs 5V to work.

- ✓ The first method is to use the USB port on the BoB and connect it to a personal computer USB female port. The only drawback on this method is that one more cable must go from the machine to the PC.
- ✓ The second method is to use a 5V power adaptor like the one that it is used to charge a mobile phone and connect it to the "+5V" and "-5V" on the breakout board.
- ✓ The third method and the one that has been used on the prototype CNC machine is to use a 220V to 5VDC converter to power up the BoB. That converter can be seen below in the photo.

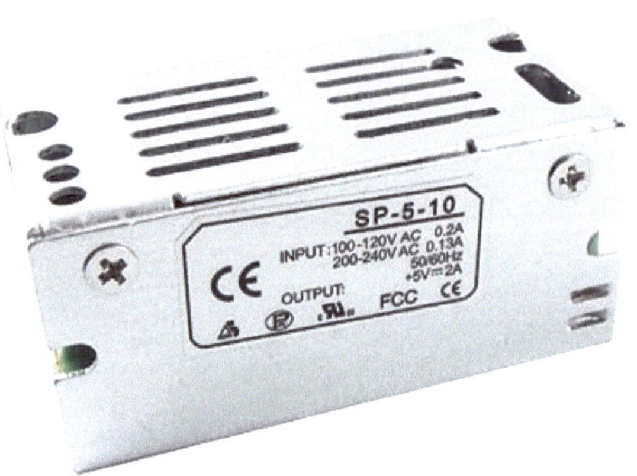

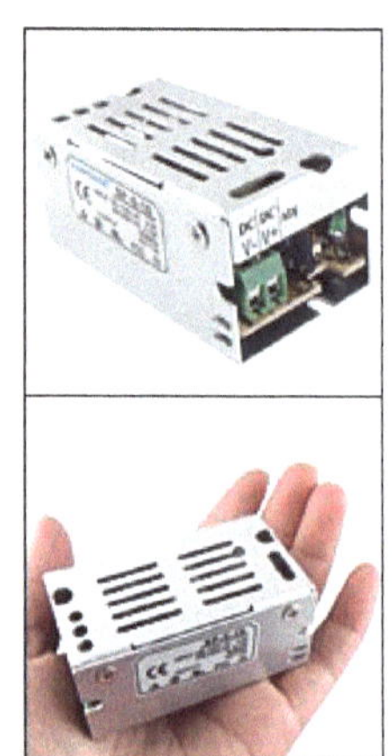

That converter connects to the 220V from the PSU and outputs adjustable voltage with an upper limit of 5VDC. It is a great device that will "save" the machine from extra cables.

Again, use a voltmeter to make sure that the output DC voltage of the converter is true 5V in order to not burn your Breakout Board.

BoB to Personal Computer (PC) – Connection

The wiring between the BoB and personal computer is done with a DB-25 Parallel Port cable used widely for printers.

Some manufactures ship a Parallel Port cable together with BoB. The only think that has to be checked is if the connection between the pins of the DB-25 cable is linked correct together inside the cable. That check can be easily done again with a voltmeter.

Stepper Motors to Drives – Connection

The stepper motors can have 4, 6 or 8 cables coming out from their core. The motors used in the prototype CNC router are 4 leads, NEMA 23, 425 oz/in motors.

The manufacturer of the stepper motors will tell you which cable is A^+, B^+, A^- and B^- in order to connect the stepper motors correct to the drives.

If you don't have any clues about the cables, you can connect any two of them and try to rotate the axis. If it's not rotating or rotating very hard then those two cables are different, one is A while the other is B. Make the possible combinations and you ll find the two pairs easily.

The connection to the drives is made easily when you connect the "A cables" to the "A ports" on the DM542A and the "B" on "B ports" respectively. If the stepper rotates in the opposite direction, just change the pairs.

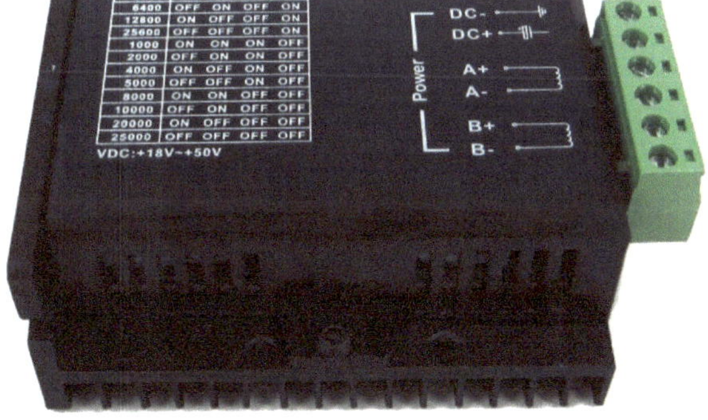

Limit Switches are the most necessary parts and should be installed in the system. These are small microswitches installed just before the end of travel on each axis. When the machine reaches this point in travel, the switch is triggered, sending a signal to the machine control software, which stops motion. This prevents the machine from accidentally slamming into the end of its travel, potentially causing damage or operator injury.

Limit switches can be installed in many wiring configurations. A compromise must be found between giving the software as much as possible information of machine position while not using an excessive number of parallel port input pins.

The diagram on the page with cable harness shows a good compromise between maximizing software knowledge while minimizing input port pins. Note that limit switches are installed as "normally closed" so that any breakage to the cable will stop the machine.

The pins on the BoB to wire each axis limit switches are listed below:

- X-Axis -> Pin 10
- Y-Axis -> Pin 11
- Z-Axis -> Pin 12

The emergency button is the most important button on the machine. It will save the machine from a collision to the machining part, to the end of an axis travel and when jogging the machine when you test it at the phase of building.

The emergency stop (e-stop) should be wired in the default input port on the breakout board with name **Pin 15**.

Also, most software programs won't let you "run" the machine without an emergency stop wired. There are some ways to bypass that feature but it is not something that I recommend.

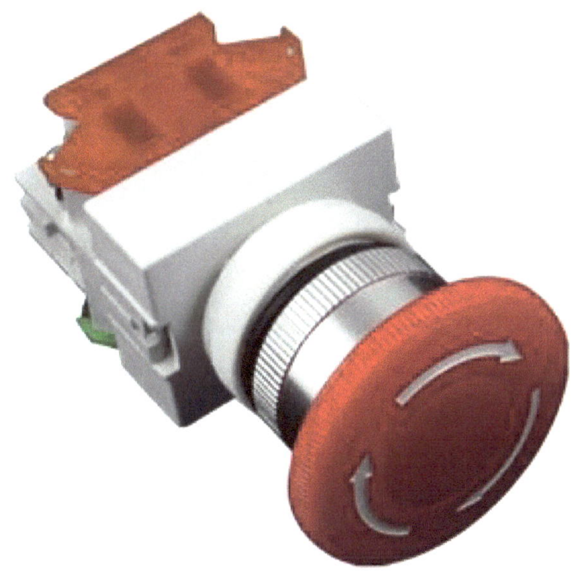

Using Mach 3 for the first time

1. Open MACH 3 software and select Mach3Mill.

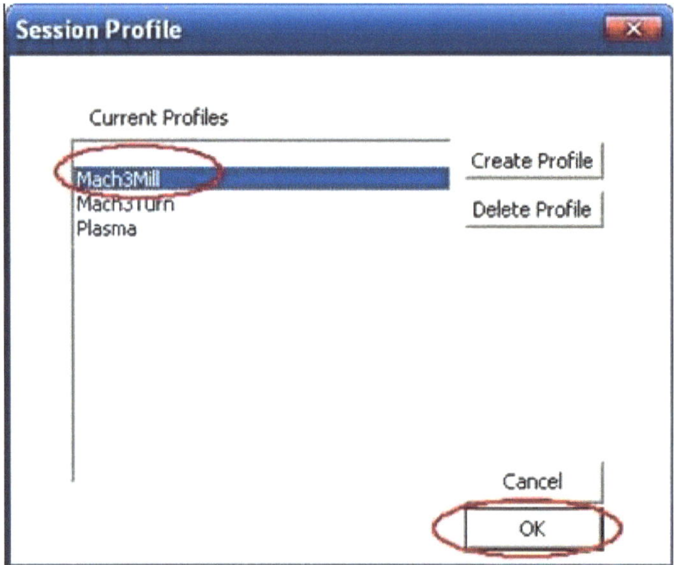

2. When the software starts, the main interface screen can be seen.

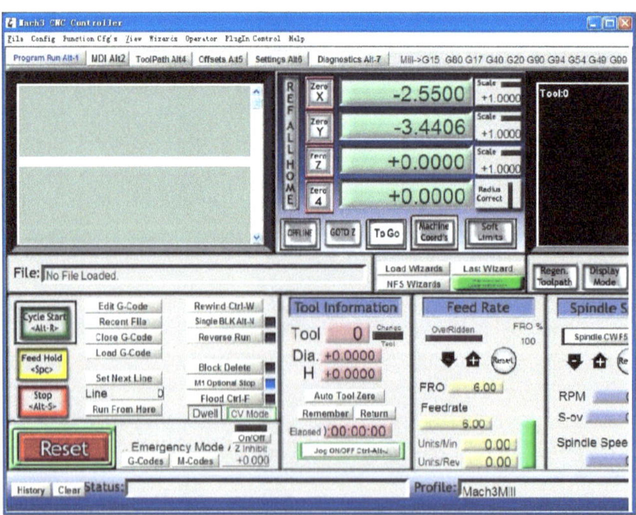

3. From the upper tab, click "Config" and then "Port and Pins".

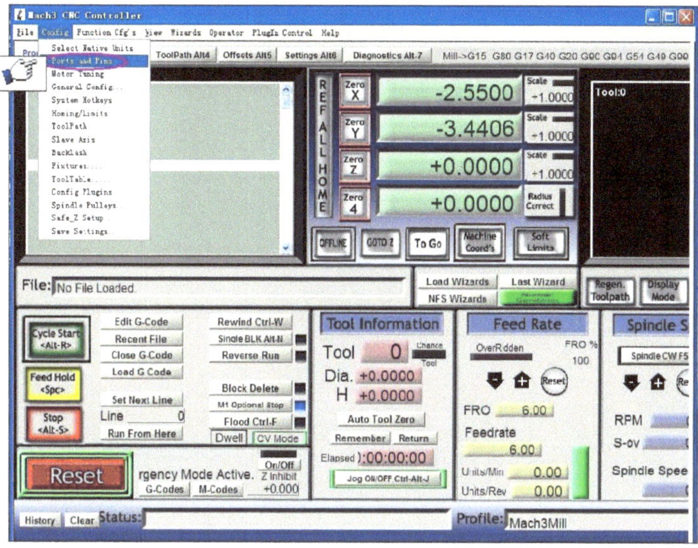

4. When the tab opens, set only the frequency (if unsure leave it on 25000Hz) and the select "Motor Outputs".

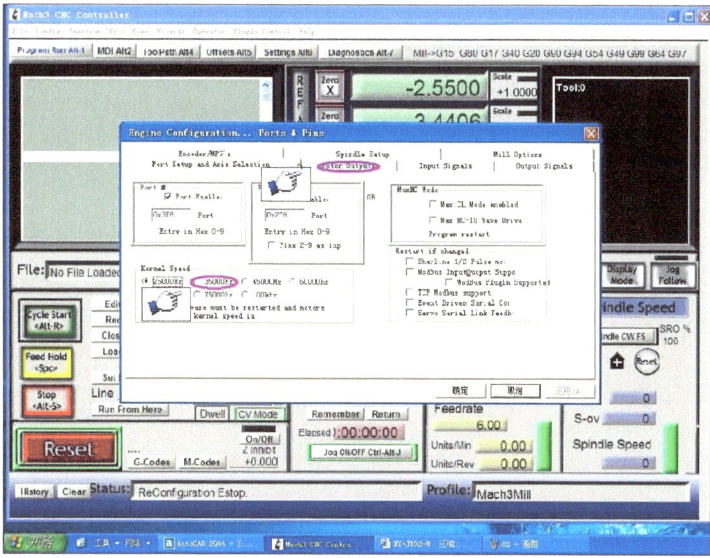

5. On the “Motor Outputs” tab, “Enable” all the Axes that you are going to use and correct the “Step” and “Dir” pins. Also, the “step” and “dir” port should be 1.

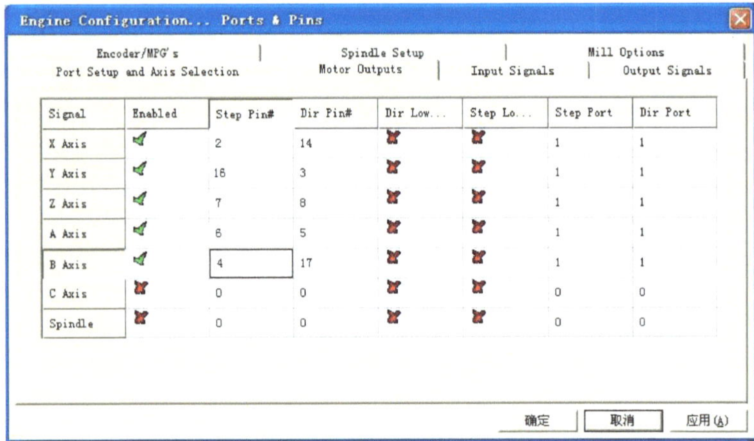

Signal	Enabled	Step Pin#	Dir Pin#	Dir Low...	Step Lo...	Step Port	Dir Port
X Axis	✓	2	14	✗	✗	1	1
Y Axis	✓	16	3	✗	✗	1	1
Z Axis	✓	7	8	✗	✗	1	1
A Axis	✓	6	5	✗	✗	1	1
B Axis	✓	4	17	✗	✗	1	1
C Axis	✗	0	0	✗	✗	0	0
Spindle	✗	0	0	✗	✗	0	0

6. On the “Output Signal” tab, “Enable” all the Axes that you are going to use and also find the “e-stop” and enable it.

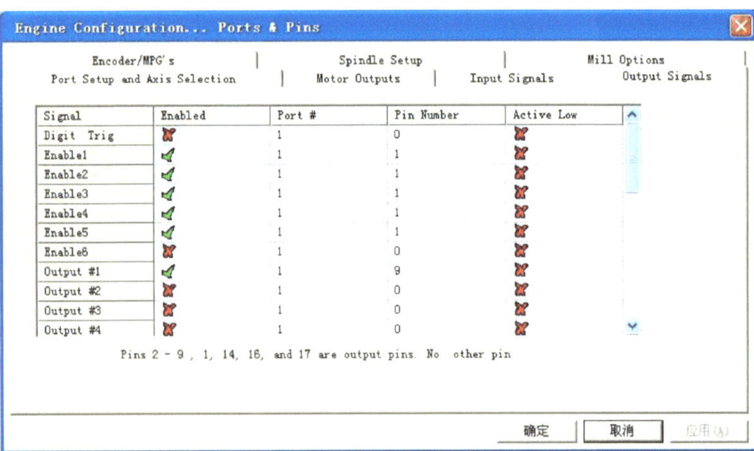

Signal	Enabled	Port #	Pin Number	Active Low
Digit Trig	✗	1	0	✗
Enable1	✓	1	1	✗
Enable2	✓	1	1	✗
Enable3	✓	1	1	✗
Enable4	✓	1	1	✗
Enable5	✓	1	1	✗
Enable6	✗	1	0	✗
Output #1	✓	1	9	✗
Output #2	✗	1	0	✗
Output #3	✗	1	0	✗
Output #4	✗	1	0	✗

7. From the upper tab, find the "Motor Tuning and Setup" tab and insert the values below for all axes, in order to test the stepper motors movement. Later with testing, those values can be changed to more suitable.

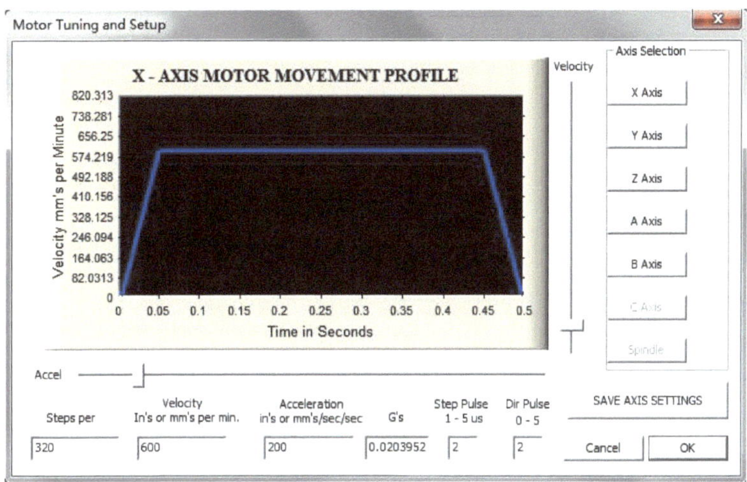

8. Now, load a new G-Code and watch your machine "dancing".

Congratulations!!!

You have completed the whole construction and made a 100% working machine. Hopefully you will have a great time using your CNC machine!

Always remember, it is a machine and machines can be dangerous!

Safety rules must be kept at all time.

"Happy cutting"

Tsipas Ilias

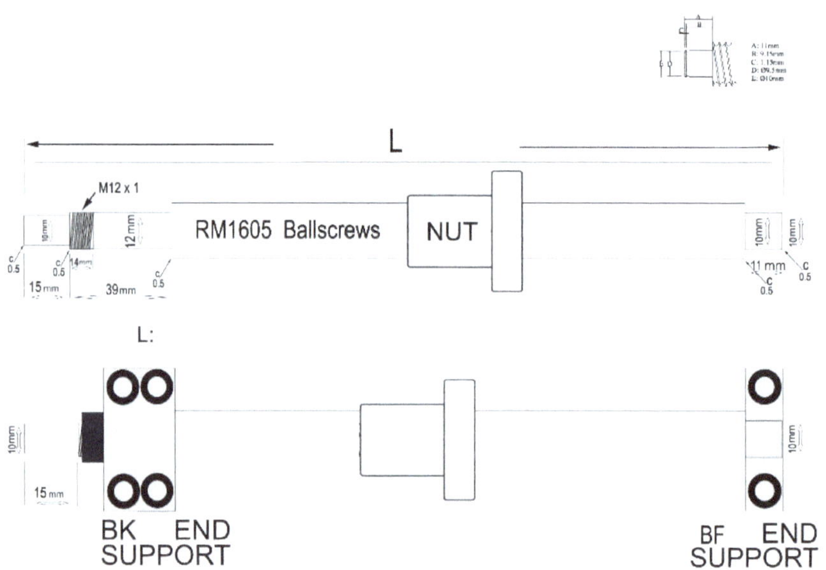

1. End-Machining in RM1605 Leadscrew

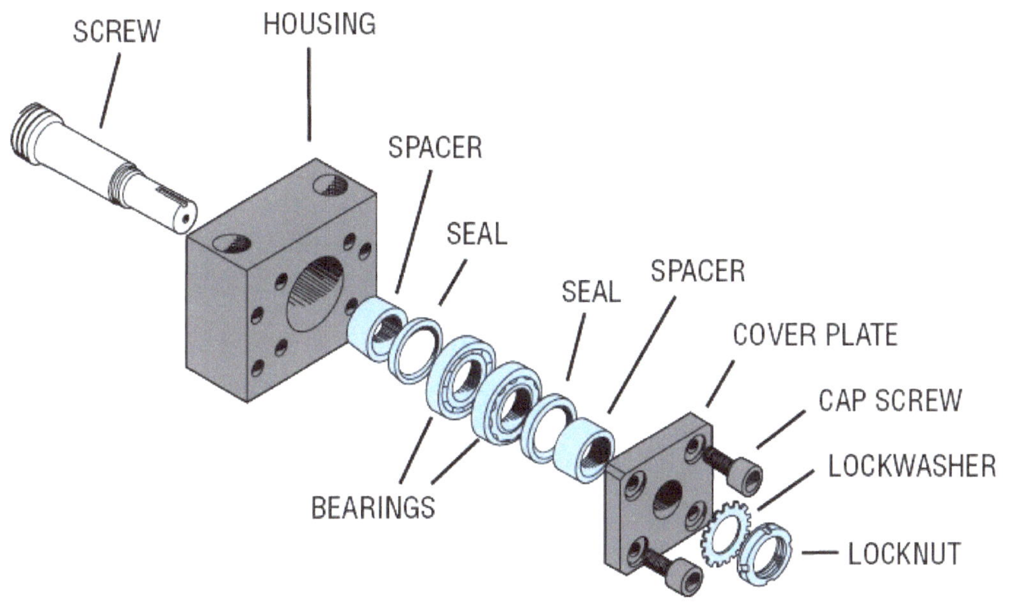

2. Method of mounting the bearings in BK Mounts

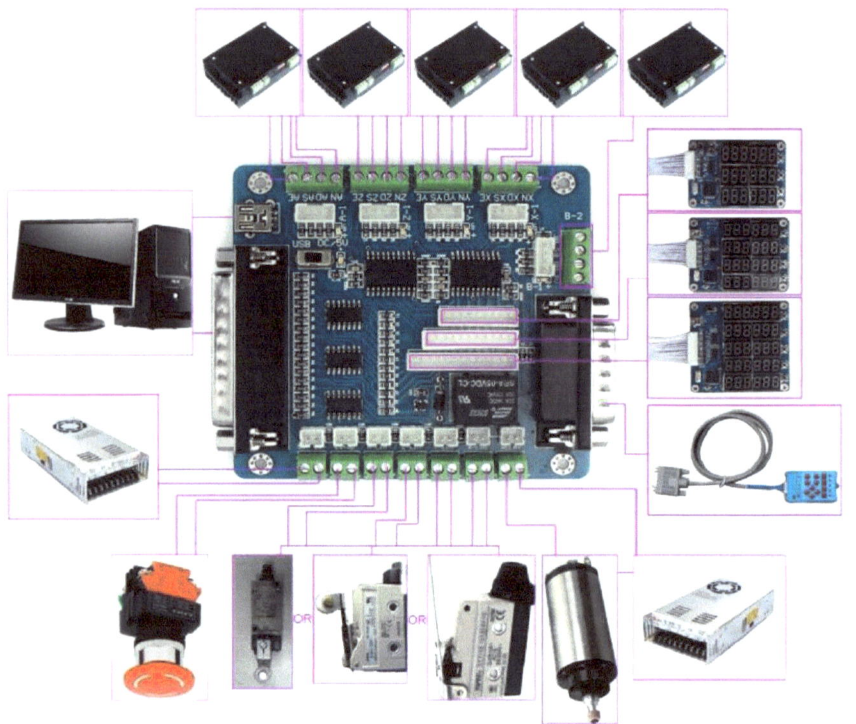

3. Connections of Breakout Board

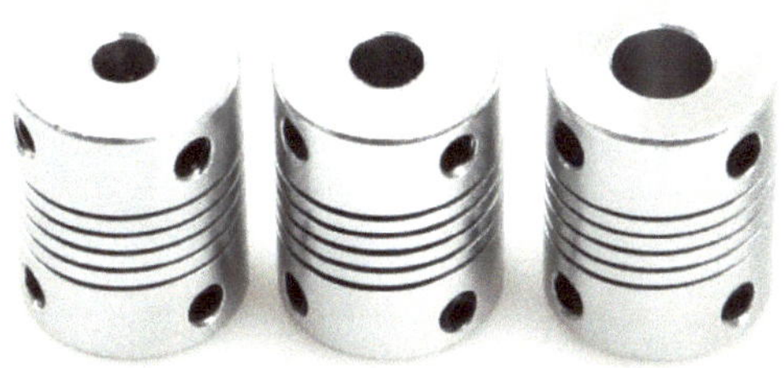

4. Typical Flexible Shaft Couplings

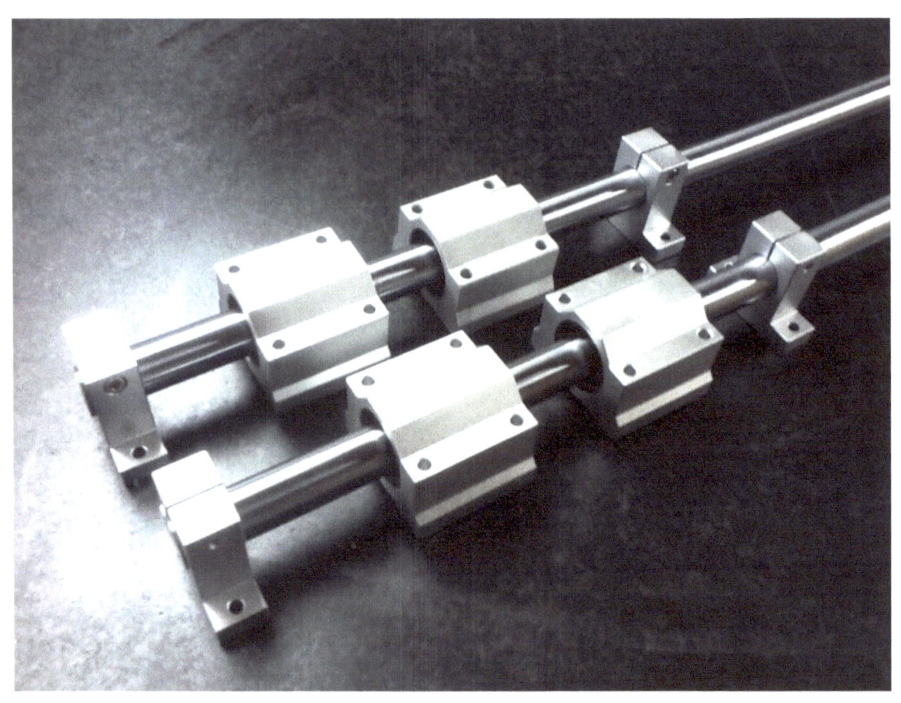

5. Typical Rails and SK Mounts

Metric-ISO threads coarse pitch				Metric-ISO threads fine pitch				Metric-ISO threads fine pitch			
M	**Pitch mm.**	**Maximun core dia. mm.**	**Drill size mm.**	**MF**	**Pitch mm.**	**Maximun core dia. mm.**	**Drill size mm.**	**MF**	**Pitch mm.**	**Maximun core dia. mm.**	**Drill size mm.**
1	0,25	0,785	0,75	**2,5**	0,35	2,221	2,15	**25**	2,00	23,210	23,00
1,1	0,25	0,885	0,85	**3**	0,35	2,271	2,65	**26**	1,50	24,676	24,50
1,2	0,25	0,985	0,95	**3,5**	0,35	3,221	3,15	**27**	1,00	26,153	26,00
1,4	0,30	1,160	1,10	**4**	0,50	3,599	3,50	**27**	1,50	25,676	25,50
1,6	0,35	1,321	1,25	**4,5**	0,50	4,099	4,00	**27**	2,00	25,210	25,00
1,7	0,35	1,346	1,30	**5**	0,50	4,599	4,50	**28**	1,00	27,153	27,00
1,8	0,35	1,521	1,45	**5,5**	0,50	5,099	5,00	**28**	1,50	26,676	26,50
2	0,40	1,679	1,60	**6**	0,75	5,378	5,20	**28**	2,00	26,210	26,00
2,2	0,45	1,838	1,75	**7**	0,75	6,378	6,20	**30**	1,00	29,153	29,00
2,3	0,40	1,920	1,90	**8**	0,75	7,378	7,20	**30**	1,50	28,676	28,50
2,5	0,45	2,138	2,05	**8**	1,00	7,153	7,00	**30**	2,00	28,210	28,00
2,6	0,45	2,176	2,10	**9**	0,75	8,378	8,20	**30**	3,00	27,252	27,00
3	0,50	2,599	2,50	**9**	1,00	8,153	8,00	**32**	1,50	30,675	30,50
3,5	0,60	3,010	2,90	**10**	0,75	9,378	9,20	**32**	2,00	30,210	30,00
4	0,70	3,422	3,30	**10**	1,00	9,153	9,00	**33**	1,50	31,676	31,50
4,5	0,75	3,878	3,70	**10**	1,25	8,912	8,80	**33**	2,00	31,210	31,00
5	0,80	4,334	4,20	**11**	0,75	10,378	10,20	**33**	3,00	30,252	30,00
6	1,00	5,153	5,00	**11**	1,00	10,153	10,00	**35**	1,50	33,676	33,50
7	1,00	6,153	6,00	**12**	1,00	11,153	11,00	**36**	1,50	34,676	34,50
8	1,25	6,912	6,80	**12**	1,25	10,912	10,80	**36**	2,00	34,210	34,00
9	1,25	7,912	7,80	**12**	1,50	10,676	10,50	**36**	3,00	33,252	33,00
10	1,50	8,676	8,50	**14**	1,00	13,153	13,00	**38**	1,50	36,676	36,50
11	1,50	9,676	9,50	**14**	1,25	12,912	12,80	**39**	1,50	37,676	37,50
12	1,75	10,441	10,20	**14**	1,50	12,676	12,50	**39**	2,00	37,210	37,00
14	2,00	12,210	12,00	**15**	1,00	14,153	14,00	**39**	3,00	36,252	36,00
16	2,00	14,210	14,00	**15**	1,50	13,676	13,50	**40**	1,50	38,676	38,50
18	2,50	15,744	15,50	**16**	1,00	15,153	15,00	**40**	2,00	38,210	38,00
20	2,50	17,744	17,50	**16**	1,50	14,676	14,50	**40**	3,00	37,252	37,00
22	2,50	19,744	19,50	**17**	1,00	16,153	16,00	**42**	1,50	40,676	40,50
24	3,00	21,252	21,00	**17**	1,50	15,676	15,50	**42**	2,00	40,210	40,00
27	3,00	24,252	24,00	**18**	1,00	17,153	17,00	**42**	3,00	39,252	39,00
30	3,50	26,771	26,50	**18**	1,50	16,676	16,50	**45**	1,50	43,676	43,50
33	3,50	29,771	29,50	**18**	2,00	16,210	16,00	**45**	2,00	43,210	43,00
36	4,00	32,270	32,00	**20**	1,00	19,153	19,00	**45**	3,00	42,252	42,00
39	4,00	35,270	35,00	**20**	1,50	18,676	18,50	**48**	1,50	46,676	46,50
42	4,50	37,799	37,50	**20**	2,00	18,210	18,00	**48**	2,00	46,210	46,00
45	4,50	40,799	40,50	**22**	1,00	21,153	21,00	**48**	3,00	45,252	45,00
48	5,00	43,297	43,00	**22**	1,50	20,676	20,50	**50**	1,50	48,676	48,50
52	5,00	47,297	47,00	**22**	2,00	20,210	20,00	**50**	2,00	48,210	48,00
56	5,50	50,796	50,50	**24**	1,00	23,153	23,00	**50**	3,00	47,252	47,00
60	5,50	54,796	54,50	**24**	1,50	22,676	22,50	**52**	1,50	50,676	50,50
64	6,00	58,305	58,00	**24**	2,00	22,210	22,00	**52**	2,00	50,210	50,00
68	6,00	62,305	62,00	**25**	1,00	24,153	24,00	**52**	3,00	49,252	49,00
				25	1,50	23,676	23,50				

6. Metric-ISO Threads – Drill and Tap Chart

Notes

References

1. Pavlik, B., (2013) *Momus Design Benchtop CNC Router Plans v.2.1.* http://www.momuscnc.com/

2. *Autodesk Inventor Professional 2015*, http://www.autodesk.com/products/inventor/overview

3. *Newfangled Solutions LLC*, ME 04254, USA, http://www.machsupport.com/

4. *Longs-Motor*, NO.18 Hengshan Road, New Area, Changzhou City, Jiangsu Province, P.R. China, http://www.longs-motor.com/,

www.ingramcontent.com/pod-product-compliance
Lightning Source LLC
Chambersburg PA
CBHW041932220326
41598CB00055BA/34

* 9 7 8 9 6 0 9 3 7 1 8 8 9 *